D0856684

Water & Wastewater Examination Manual

Water & Wastewater Examination Manual

V. Dean Adams

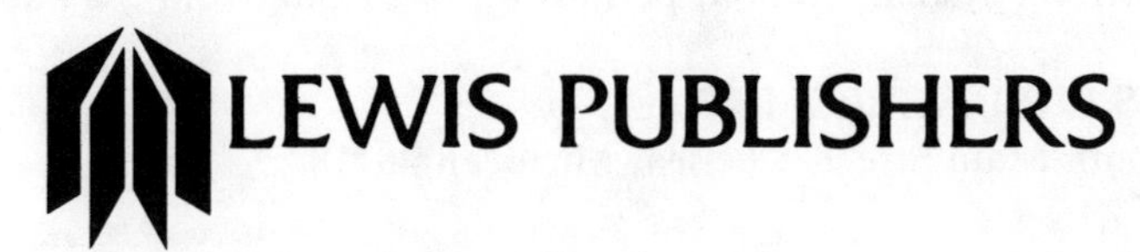

Library of Congress Cataloging-in-Publication Data

Adams, V. Dean.
Water and wastewater examination manual / by V. Dean Adams.
p. cm.
Includes bibliographical references.
1. Water—Analysis—Laboratory manuals. 2. Sewage—Analysis—Laboratory manuals. I. Title.
TD380.A33 1990
628.1'61—dc20 90-5848
ISBN 0-87371-199-8

LEWIS PUBLISHERS, INC.
121 South Main Street, Chelsea, Michigan 48118

PRINTED IN THE UNITED STATES OF AMERICA

Acknowledgments

Many individuals have contributed to the content of this manual. Thanks are due to Dr. Peter Cowan, Dr. Don Porcella, Dr. Mary Cleave, Ms. Leslie Terry, Ms. Mary Nubbe, Ms. Alberta Seierstad, and Mr. Jeff Curtis, all of whom, over the past several years, have made numerous suggestions for its improvement. Several laboratory technicians and students have also contributed substantially by verifying the procedures described in the book. I also thank Ms. Bobbi Nabors and Ms. Patricia LaFever for typing the manuscript.

Special thanks go to my wife, Joyce, our sons, Todd, Doug, and Ned, and our families who have assisted me in my professional development. This effort would not have been possible without their dedication, help, and encouragement.

Preface

The *Water and Wastewater Examination Manual* was written using newly revised procedures from *Standard Methods for the Examination of Water and Wastewater* (1989), *Microbial Methods for Monitoring the Environment: Water and Wastes*, *Methods for Chemical Analysis of Water and Wastes* (EPA, 1978, 1983), and other pertinent references. It is intended to be used as a working laboratory guide and reference to water quality analysis for upper level undergraduates, graduate students, water quality technicians, and water and wastewater plant operators. Procedures are included for parameters that are frequently needed in water quality analysis. In general, the analytical procedures have been thoroughly examined and tested in the Water Quality Laboratory at the Utah Water Research Laboratory, Utah State University, Logan, Utah and the Water Quality Laboratory at the Center for the Management, Utilization, and Protection of Water Resources, Tennessee Technological University, Cookeville, Tennessee. The procedures in this manual were selected from many available methods; the tests included here are those that are commonly used in water and wastewater laboratories. Consult the Bibliography for additional information on other techniques.

This manual is not intended to be a replacement for *Standard Methods* or for the EPA manuals on water quality analysis. In many cases, *Standard Methods* and the EPA manuals give detailed information about such aspects of the analytical procedures as interferences, not included here. However, many sections are expanded compared to *Standard Methods*, with instructions on preparing standards, reagent additions, preparation of spiked samples, and special techniques where applicable. A few reliable and valuable procedures are also included here that are not listed in *Standard Methods*.

V. Dean Adams earned a BS degree in chemistry from Idaho State University and then studied with Dr. Richard C. Anderson at Utah State University, receiving his PhD in chemistry in 1972. The next thirteen years were spent at the Utah Water Research Laboratory and the Division of Environmental Engineering at Utah State University, where he served as the supervisor of the Environmental Water Quality Laboratory and the head of the Division, in addition to teaching and research responsibilities.

In 1985, Dr. Adams was appointed the Director of the Center for the Management, Utilization, and Protection of Water Resources at Tennessee Technological University, where he is also a professor in the Departments of Civil Engineering and Chemistry. Under his direction, the Center was awarded the 1987 Performance Par Excellence Award by the Tennessee Higher Education Commission and, in 1988, was named an Accomplished Center of Excellence.

Dr. Adams has been active in the Division of Environmental Chemistry of the American Chemical Society, having served as the secretary, chairman-elect, and chairman. He also serves on the Advisory Boards of the American Chemical Society Books Department and the CHEMTECH magazine.

Dr. Adams' research interests span a wide variety of water and environmental issues, including the fate of contaminants in the environment; photo-enhanced degradation of hazardous wastes; environmental evaluation and management of lake systems; and the use of bioassays in the assessment of impacts of environmental contaminants. He has authored or co-authored over 115 refereed research papers, technical reports, and book chapters on environmental chemistry and engineering. He also holds two wastewater patents.

Table of Contents

4. Methods for the Determination of Inorganics and Nonmetals

List of Tables

List of Figures

Introductory Guidelines for the Use of This Manual

1. Introductory Guidelines for the Use of This Manual

1.1 Safety
(Adapted from *Standard Methods*, 1989)

A. General

Safety is a subject of concern for all. In the specific context of laboratories, it is especially important because of the presence and use of hazardous chemicals or equipment, microorganisms that may produce human disease, and radioactive substances. Laboratory management has a vital role to play in ensuring a safe working environment. This role is dictated by federal or state regulations as well as by the appropriate interests of management in the well-being of its employees and the overall productivity of the laboratory. In the final analysis, however, safety is the responsibility of the individual laboratory worker. While management must deal with the work place, the provision of safe procedures, and the training of employees in safe practices, it is the analyst who ultimately makes the laboratory safe or unsafe.

B. Laboratory Hazards

Laboratory hazards fall into two major categories: those of universal significance, such as fire, electrical, and mechanical hazards, and those specific to a laboratory, such as chemical or biological hazards. Because this is not intended as a manual of safe practices, general safety problems will not be discussed. However, the use of combustible or explosive reagents and the widespread use of electrical equipment may pose special problems in the laboratory.

BE SURE TO READ A COMPREHENSIVE SAFETY MANUAL BEFORE WORKING IN THE LABORATORY.

From the standpoint of chemical and biological safety, it must be recognized that many reagents require handling with the utmost care either in their original state or in solution, or both. This care is exercised either to protect the health and safety of the analyst or because of environmental hazard or damage, particularly in connection with laboratory waste disposal. A number of reagents specified in this manual may bear labels with the words **POISON, DANGER, CAUTION, FLAMMABLE**, etc. Handle these with special care.

The safest way to deal with hazardous materials is to avoid their use. Efforts to eliminate such reagents were made in selecting analytical methods for inclusion in the 1989 edition of *Standard Methods* and in this manual. When avoidance is impossible, hazardous materials are identified, and special precautions against exposure by inhalation, by ingestion, or through the skin are given.

Information on laboratory safety and chemical disposal is available. Reference materials include (see Bibliography): American Water Works Association (1958); Inhorn (1978); Manufacturing Chemists' Association (1972); Phifer and McTigue (1988); Bennett et al., 1982; Walters (1980); Dux and Stalzer (1988); *Safe Storage of Laboratory Chemicals* (1984); National Research Council (1981, 1983); *Hazards in the Chemical Laboratory* (1983); and *Safe Handling of Cryogenic Liquids* (1987).

C. Safety Equipment

Simple safety practices include using mechanical pipettors instead of pipetting by mouth and using safety clothing and fume hoods. Safety glasses or, still better, full face masks offer important protection against explosions or implosions. Additional safety equipment includes fire extinguishers, fire blankets, safety showers, eye wash fountains, and first aid and spill control kits. All laboratory personnel and students working in the lab should know the location of these safety devices and be familiar with their use.

If not sure . . . ASK!

D. Hazardous Materials in this Manual

1. *Acids and Alkalies*: Concentrated acids and bases may cause chemical burns and are especially hazardous if spilled or splashed into the eyes. Always handle with extreme care to avoid contact. In diluting concentrated acids, always add the acid to water to prevent possible explosion; *never add water to the acid.*

2. *Arsenic*: Inorganic arsenic compounds are used to prepare standards and may be present in samples. Arsenic is highly toxic and may cause lung cancer or death: avoid inhalation, ingestion, and skin exposure. Always make dilutions in a hood.

3. *Azides*: Sodium azide is used in a number of procedures including the test for dissolved oxygen. It is toxic and reacts with acid to produce the still more toxic hydrazoic acid. When discharged to a drain, it may react with and

accumulate on copper or lead plumbing fixtures. The metal azides are explosive and detonate readily. Avoid inhalation, ingestion, and skin exposure. Destroy azides by adding a concentrated solution of sodium nitrite, $NaNO_2$ (1.5 g $NaNO_2$/g sodium azide). To remove accumulated metal azides from drainpipes and traps, treat overnight with a 10% solution of sodium hydroxide.

4. *Biohazards*: Samples may contain pathogenic microorganisms. Exposure to these organisms may be incidental to chemical or biological examination or occur in the specific examination for certain disease-producing organisms. In either case, avoid ingestion particularly in culturing pathogens. Use aseptic techniques and sterilize all discarded cultures.

5. *Compressed Gases*: Compressed gases are used widely in most laboratories, especially if an atomic absorption spectrophotometer or a gas chromatograph is used. The gases may be flammable or explosive and require careful handling. Protect the cylinders themselves from freezing, overheating, and mechanical damage. Chain, lock, or otherwise prevent the cylinders from moving or falling over. Use the appropriate pressure-reducing valve for each type of gas cylinder.

6. *Cyanides*: Cyanides are used as reagents or may be present in samples. Most cyanides are toxic; avoid ingestion. Handle such solutions in a fume hood and avoid inhalation. In acid solution, the toxic gas hydrogen cyanide may be produced; therefore, do not acidify cyanide solutions.

7. *Mercury*: Mercury and its compounds are used to prepare standards, displace gases, serve as indicator liquid in thermometers, and preserve samples. Liquid mercury is a toxic volatile element. Handle spills expeditiously to prevent inhalation. Keep powdered sulfur on hand to spread immediately on mercury spills to minimize volatilization before cleanup. Disposal of samples containing mercury is environmentally damaging. Store mercury waste solutions in labeled containers. The mercury will be precipitated and collected by filtration for disposal.

8. *Perchloric Acid*: Perchloric acid is used in digesting organic matter. It can react explosively with organic matter and must be handled with care; predigest samples containing organic matter with nitric acid before adding perchloric acid, and do not add perchloric acid to a hot solution. Like azides in a drain, perchlorates may accumulate in a hood or air exhaust system. Accumulated perchlorates may react explosively with organic matter; *use special perchloric acid fume hoods and ducting when using perchloric acid digestion procedures*.

9. *Toxic or Carcinogenic Organic Compounds*: Organic solvents and solid organic reagents are used in many determinations. These may be flammable or explosive, and as such require special handling and storage, or they may be toxic or carcinogenic. Handle such solvents as chloroform, carbon tetrachlo-

ride, benzene, etc., in a fume hood and avoid inhalation, skin contact, and ingestion.

Note: In general, most waste organic solvents can be repurified by distillation and reused. Store in labeled waste containers. Distill carefully in a distillation apparatus where the temperature is maintained at the boiling point of the solvent in question.

1.2 Quality Control and Quality Assurance

A. References

See *Standard Methods* (1989, pp. 1-1 through 1-30), EPA (1979), *Quality Assurance Manual* (1979), and Taylor (1987).

B. Introduction

A good quality control program at an analytical laboratory involves a number of practices. It goes beyond simply having adequate personnel, facilities, equipment, reagents, and standards. On a routine basis, quality control samples must be analyzed along with unknown samples (see 1.2 C). Many laboratories also undergo routine proficiency testing by analyzing independent reference samples for laboratory evaluation (external quality control). In addition, each laboratory should maintain precision and accuracy data for each analysis (see 1.2 D, 1.3, 1.4, and 1.5).

C. Internal Quality Control

Internal quality control samples are generally run on a quarterly basis. These samples are available from the EPA Environmental Monitoring and Support Laboratory and from certain independent supply companies. The "true" values for these samples are usually the result of many analytical determinations and the analyses are often run by several different laboratories.

The analysis of a quality control sample serves as an evaluation of a technician's or student's analytical technique. Acceptable quality control results must be maintained for all analyses completed each quarter. Internal quality control samples are available from the laboratory's quality assurance chemist. Indications will be given as to the approximate concentration range (s) expected. These unknowns should be run in conjunction with the normal sampling and analysis schedule.

D. Precision and Accuracy

Precision is a measure of the reproducibility of a method when it is repeated on a homogeneous sample under controlled conditions. When a series of replicate analyses are done and the results are handled mathematically, if the experimental values are very close together, then the work shows good precision. However, precise data do not necessarily imply that the data are also accurate. Sometimes a set of precise observed values can be widely different from the true value because of systematic or constant errors inherent in the analytical methods or due to sample characteristics. Precision can be expressed mathematically as standard deviation or relative standard deviation.

Accuracy is a measure of the departure of a measurement from the true or actual value. It is determined by the addition of a known quantity of chemical standard to a sample and subsequent analysis to determine the percent recovery of that known quantity. Accuracy can be expressed mathematically as percent recovery or relative error.

1.3 Precision; Precision Statements

On a yearly basis, or when a new method is being adopted, precision data for actual samples must be compiled. Precision can often vary with the concentration level of the samples. For this reason, four concentration levels are studied: (1) low level—choose a sample with a concentration near the detection limit of the method, (2) intermediate level—choose two samples here, and (3) high level—choose a sample with a concentration near the upper limit of application of the method. Refer to Tables 1.1 through 1.8 (Tables 1.1–1.19 are located at the end of Chapter 1) for recommendations for the concentration levels for each method. Note that the concentration ranges pertain to the applicable range of the analysis, not to the possible concentration of a sample that may be encountered in the laboratory.

Perform a minimum of seven replicate analyses for the parameter evaluated using the four levels chosen above. Table 1.9 provides a form useful in recording this data.

It is desirable to allow the maximum interferences due to sequential operation. Therefore, run the samples in the following order: high, low, intermediate, intermediate, etc., in series seven or more times. In other words, do not read all replicates of one sample in one sequence.

To allow also for maximum changes in instrument operation, it is desirable to carry out this precision study over at least 2 hr of normal laboratory operation. Using the data collected, calculate the mean ($\bar{x}$), standard deviation (s), and relative standard deviation (RSD) as shown in Tables 1.9 and 1.10.

The precision statement, as shown at the bottom of Table 1.9, is for filing purposes and summarizes the statistical data over the entire concentration range studied. It utilizes only the data for the high and low extremes of the range.

Refer to Accuracy; Accuracy Statements (1.5). It is desirable to obtain accuracy and precision data on the same samples on the same day.

Completed precision data and statements should be filed with those previously determined. Check for significant changes in the relative standard deviation with time. For each concentration range, compile an average relative standard deviation using data from previous years. Daily replicate data (see 1.4) should agree with the accumulated yearly precision statements.

1.4 Precision; Daily Replicate Sample Analyses

On a daily basis, duplicate analyses must be performed on a minimum of 10% of the samples in an effort to ensure reproducible results. Choose the samples for replicate analyses in a random, unbiased manner. If fewer than ten samples are being analyzed, do a duplicate determination on one sample. If the results for any particular set of samples consistently fall at or below the detection limit of a test, then perform a duplicate analysis on an internal quality control.

Refer to the section on spiked samples in Accuracy; Accuracy Statements (1.5). The same samples chosen above should be carried through the spiking procedure concurrently.

All daily replicate data should agree with precision data that are generated and filed on a yearly basis. See Table 1.11 for replicate sample analyses for ammonia, NH_3-N.

For each sample, compare the percent deviation from the average to the RSD from compiled precision data for that particular concentration level. (For the purposes of this discussion, the precision data in Table 1.9 are treated as compiled data). Replicate data are *acceptable* when the percent deviation is *within ±2 relative standard deviations* from the mean for that concentration level. The analysis is considered to be at the warning level when the percent deviation is *between ±2 and ±3 relative standard deviations* from the mean. The results are considered to be *not acceptable* when the percent deviation is greater than *±3 relative standard deviations* from the mean.

In Table 1.11, the replicate analyses for sample 6309–88 are not acceptable since the percent deviation exceeds ±3 RSD (±5.7%) for that concentration level. The replicates for sample 6319–88 are at the warning level since the percent deviation falls between ±2 (±28%) and ±3 (±42%) RSD. Replicate analyses for 6329–88 and 6339–88 are acceptable.

When a result falls outside the control limits, steps must be taken to determine the cause. First, determine whether a calculation error was made. Then check to see if the instruments used for the analysis are functioning properly. Then reanalyze the sample(s) and standards in question. If these steps do not bring the analysis back within acceptable control limits, then the sample(s) must be prepared again and analyzed. It may be necessary at this time to prepare the standards again also. If all of the above procedures do not bring the analysis under control, then notify the laboratory supervisor. He or she will make the final decision as to whether the data will be edited, deleted, or handled in some other manner. Other sample data collected concurrently with unacceptable precision data are generally considered unreliable.

It is important to note that until a substantial amount of precision data are compiled, the control limits defined above may not be totally justified or correct. Also, occasions may arise when large numbers of results fall outside the control limits. This may indicate that everyday precision in the laboratory cannot be as good for certain analyses as the precision statement indicates.

As an alternative to the comparison of results to the compiled precision data, sample control charts may be constructed. Consult the references for examples and discussions of these charts.

If there are no compiled precision data available, it is still desirable to check sample replication. In this case, check the references listed for individual procedures in this manual. Precision and accuracy statements are often listed at the end of the method. It is often necessary to calculate relative standard deviations from the standard deviation data given here. Apply the same criteria as discussed above to determine whether the daily replicate data are acceptable, at the warning level, or not acceptable.

1.5 Accuracy; Accuracy Statements

Accuracy data are compiled on a yearly basis or when a new method is being evaluated. Precision data should be compiled at the same time on the same samples. Accuracy is determined by adding a known amount of the constituent analyzed to the sample, producing what is known as a spiked sample. Accuracy is reported as the percent recovery of the spike.

As with precision, accuracy can vary with the concentration level of the samples. Refer to Precision; Precision Statements (1.3) and choose four samples in the manner outlined there. Again, a minimum of seven replicate spikes is required for reliable compilation of accuracy (percent recovery) data.

Tables 1.1 through 1.8 outline spiking procedures for most of the analyses in this manual and the equations for calculating the theoretical concentration of a spiked sample (TH-SP). Note that TH-SP equations differ with the analysis in question and the spiking procedure used. These tables are intended to give suggestions for spiking samples; spiking solutions and procedures can be changed as the analyst sees fit, but TH-SP calculations must be modified accordingly. Note also that the concentration ranges for each level (high, intermediate, and low) pertain to the applicable range of the analysis in question, not to the possible concentration of an actual sample encountered in the laboratory. Spiking solutions and compounds for the high concentration levels are less concentrated than those for samples falling in the intermediate ranges. This spiking design should minimize the instances when the concentration of a spiked sample exceeds the upper limit of the range for that analysis. Make certain that all spiking solutions or compounds are prepared accurately and carefully (see 1.9, 1.10).

If desired, the spiking solution or compound can be added to a known volume of doubly deionized water (DDW) (in place of the sample) and carried through the analytical procedure also. (In this case, UNSP in the TH-SP equation becomes zero or the value determined for the blank). Percent recovery of this spike should be very close to 100%.

As with the collection of precision data, it is desirable to allow maximum interferences due to sequential operation. Run the samples in the following order: high, low, intermediate, intermediate, etc., in series seven or more times. Do not read all replicates of one sample in one sequence.

Allow also for maximum changes in instrument operation by carrying out this accuracy study over at least 2 hr of normal laboratory operations.

Enter data from the replicated, spiked samples on a form similar to those shown in Tables 1.12 and 1.13. Using the calculations outlined in Table 1.10, compare the theoretical concentration of the spike (TH-SP) to the experimentally determined concentration of the spiked sample (EX-SP) to determine the percent recovery of the spike (%R) for each spiked sample replicate. Then calculate the mean ($\overline{\%R}$), standard deviation (s), and relative standard deviation (RSD) as outlined in Table 1.10.

The accuracy statements at the bottom of Tables 1.12 and 1.13 are for filing purposes and summarize the statistical percent recovery data over the entire concentration range studied. It utilizes only the data for the high and low extremes of the range.

Completed accuracy data and statements should be filed with those previously determined. Check for significant changes in the relative standard deviation and percent recovery with time. For each concentration range, compile an average relative standard deviation using data from previous years. Daily spiking data should agree with the accumulated yearly accuracy statements.

1.6 Accuracy; Daily Spiking Sample Analyses

On a daily basis, spiking analyses must be performed on a minimum of 10% of the samples in an effort to ensure accurate results. Choose the samples for spiking analysis in a random, unbiased manner. If fewer than 10 samples are being analyzed, spike one sample. If the results for any particular set of samples consistently fall at or below the detection limit of a test, then spike an internal quality control.

Refer to Precision; Daily Replicate Sample Analyses (1.4). The same samples chosen above should be carried through the precision procedure concurrently.

All daily spiking data should agree with accuracy data that are generated and filed on a yearly basis. See Table 1.14.

For each sample, compare the percent deviation to the RSD from compiled accuracy data for that particular concentration level. (For the purposes of this discussion, the accuracy data in Table 1.12 are treated as compiled data). Spiking data are acceptable when the percent deviation is *within ±2 relative standard deviations* from the mean for that concentration level. The analysis is considered to be at the *warning level* when the percent deviation is *between ±2 and ±3 relative standard deviations* from the mean.

In Table 1.14, the spiking analysis for sample number 6309-88 is not acceptable since the percent deviation exceeds ±3 RSD (±6.3%) for that concentration level. The spiking analysis for sample 6329-88 is at the warning level since the percent deviation falls between ±2 (±3.4) and ±3 (±5.1) RSD. Spiking analyses for 6319-88 and 6339-88 are acceptable.

When a result falls outside the control limits, steps must be taken to determine the cause. First, determine whether a calculation error was made. Then check to see if the instruments used for the analysis are functioning properly. Then reanalyze the sample(s) and standards in question. If these steps do not bring the analysis back within acceptable control limits, then the spiked sample must be prepared again and analyzed. It may be necessary at this time to prepare the standards again also. If all of the above procedures do not bring the analysis under control, then notify the laboratory supervisor. He or she will make the final decision as to whether the data will be edited, deleted, or handled in some other manner. Other sample data collected concurrently with unacceptable spiking data are generally considered unreliable.

It is important to note that until a substantial amount of accuracy data are compiled, the control limits defined above may not be totally justified or correct. Also, occasions may arise when large numbers of results fall outside the control limits. This may indicate that everyday accuracy in the laboratory cannot be as good for certain analyses as the accuracy statements indicate.

As an alternative to the comparison of results to the compiled accuracy data, sample control charts may be constructed. Consult the references for examples and discussion of these charts.

If there are no compiled accuracy data available, it is still desirable to check sample accuracy by spiking. In this case, check the references listed for individual procedures in this manual. Precision and accuracy statements are often listed at the end of the method. It is often necessary to calculate relative standard deviations from the standard deviation data given here. Apply the same criteria as discussed above to determine whether the daily spiking data are acceptable, at the warning level, or not acceptable.

1.7 Sample Preservation

Once the sample has been collected, it should be analyzed as quickly as possible or preserved and stored in a container in an attempt to maintain the sample integrity. Complete preservation for every component in the sample is essentially impossible. The preservation procedures are an attempt to modify (retard or stop) the chemical and biological changes that can occur after the sample has been collected and placed in a storage container. Approved methods of preservation are pH control, chemical addition, refrigeration, and freezing in an attempt to (1) slow biological reactions, (2) control chemical reactions (hydrolysis, oxidation, reduction, etc.), (3) reduce volatility, and (4) reduce adsorption of the components of interest.

It is best to analyze the samples as soon as possible after collection, but if immediate analysis is not possible to maintain the integrity of a sample, appropriate pretreatment, container selection, and holding times are fundamental to sample preservation.

The recommended preservative for various constituents is given in Table 1.15 from EPA (1979, 1982, 1983), *Standard Methods* (1989), and the EPA Region 8 Quality Assurance Coordinator's Meeting (July 22–23, 1981, Denver, Colorado).

1.8 Suggestions for Filtering of Water Samples

A. General

This section is a *guideline* for filtering samples. Depending on the origin and composition of the water sample and on the analytical information of interest, one may wish to deviate from Table 1.16. For example, a drinking water sample may not require any filtering. At the other extreme, a sample from an eroding stream bed may require filtering before most of the analytical parameters are run.

B. Dissolved Constituents; Types of Filters

Standard Methods (1989) defines "dissolved" as that which passes through a 0.45 μm filter. Hence, for analyses designated as dissolved (dissolved metals, dissolved organic carbon, dissolved BOD, etc.) use sample filtered with a 0.45 μm glass fiber or membrane filter; Table 1.16 lists the type of filter to use.

1.9 Use of Volumetric Glassware

Volumetric glassware is calibrated glassware used for precise measurements of volume. This includes volumetric flasks, volumetric pipets, and calibrated burets. The discussion here is limited to the use of volumetric flasks and pipets. This volumetric glassware is used in preparing reagents, stock standards, standard solutions used daily, and any necessary dilutions of samples. Less accurate measuring glassware includes graduated cylinders and serological pipets; these are used when exact volumes are unnecessary. When using this manual, always select volumetric glassware when the volumes are specified to a tenth of a milliliter. For example, use volumetric glassware for a volume listed as 50.0 mL; for a volume of 50 mL, volumetric glassware may be used but is not required. In general, volumetric flasks are used for mixing solutions, since they are calibrated "To Contain" (TC), and pipets are used for accurately measuring volumes to be transferred, since they are calibrated "To Deliver" (TD). (See Definitions of Abbreviations, Terms, and Units at the back of this book.)

Volumetric glassware must be used correctly in order to obtain accurate results. The bottom of the meniscus of the liquid must be at the line that is marked on the flask or pipet. The glassware and the liquid must be at room temperature; the glassware is usually calibrated for liquids at 20°C. One liter of water will increase in volume by about 0.2 mL for every 1°C. For volumetric flasks, insert the ground glass stopper and twist it slightly to ensure a good fit. Mix the solution by inverting the flask many times. For volumetric pipets, first rinse the pipet three times with the solution to be transferred. Then fill to the calibration mark and drain by holding the pipet in a vertical position. Allow the pipet to drain by gravity; do not blow into the pipet to hurry the process. Touch the tip of the pipet to the wall of the container for a few seconds after the liquid has drained. Do not remove the small amount of liquid in the tip; this quantity is taken into account in the calibration of the pipet.

1.10 Making Standard Solutions

A. Stock Standards

1. Preparation of the reagent grade chemicals. Follow the instructions given for the specific test. In general, if an "anhydrous" chemical is to be oven dried, place the dry chemical in a clean, dry beaker and cover. Dry at the specified time and temperature (usually a few hours at 103°C). Cool the chemical in a desiccator until it is at room temperature, 20°C, before weighing.

2. Weigh accurately. Always use an analytical balance capable of weighing to the nearest 0.1 mg. Zero the balance and then use weighing papers. Either tare the paper, or add the paper's weight to the amount of standard to be weighed. Weigh the chemical accurately.

3. Transfer the chemical quantitatively. In a pre-rinsed volumetric flask of the proper size, add some fresh DDW. Quantitatively transfer the chemical that was weighed into the flask, rinsing the paper if necessary. Add more fresh DDW, being sure to rinse any dry chemical into the bottom of the flask. Add DDW so that the flask is about three quarters full. Carefully swirl or stir the solution until the chemical is dissolved. Fill the volumetric flask with DDW so that the meniscus of the solution is at the calibration line. With a ground glass stopper in place, invert the solution several times to mix thoroughly.

4. Store in the proper container. This is usually high quality glass with a glass stopper. Follow any particular instructions given for the test, (such as using a dark container, etc). Stock standards are usually refrigerated.

B. Daily Procedure

1. Room temperature should be ~20°C. Always allow a stock standard to come to room temperature before pipeting out solutions.

2. Rinse volumetric pipets. Always use volumetric pipets for standard preparation and always rinse the pipet with the solution three times before pipeting the specified amount.

3. Use volumetric flasks. Use volumetric flasks for diluting amounts of stock solution up to a particular concentration. For all analyses, rinse the volumetric flask with DDW prior to pipeting the stock solution. For nitrogen and phosphorus analysis, rinse the volumetric flask first with 6 N HCl and then with DDW.

When diluting, add DDW so that the meniscus of the solution is at the calibration mark on the volumetric flask. Insert the ground glass stopper and twist the stopper slightly to obtain a secure fit. Mix the solution by inverting several times.

1.11 Preparation of Acid and Alkaline Solutions

A. Acidic Solutions

Prepare dilute acid solutions by cautiously adding the required amount of concentrated acid (see Table 1.17) in a hood with stirring to about 400 mL of DDW. Cool and dilute to 1 L.

Note: Always add acid to water; never add water to concentrated acid. Some reactions may be violent. Wear protective clothing and safety glasses.

B. Alkaline Solutions

1. *Sodium Hydroxide, Potassium Hydroxide*: Prepare dilute alkaline solutions by boiling about 600 mg of DDW for a few minutes to expel CO_2 gas. Cover the container and allow to cool. Cautiously add the weight of NaOH or KOH indicated in Table 1.18 with stirring in a hood. When the solute is dissolved, cool and dilute to 1 L. Store in polyethylene bottles (rigid, heavy-type) with polyethylene screw caps.

To prevent against changes in normality with time, it may be necessary to attach a tube filled with CO_2-absorbing material (soda lime, Ascarite, etc.) to the solution bottle. Withdraw solution by a siphon to avoid opening the bottle.

2. *Ammonium Hydroxide; NH_4OH*: Ammonium hydroxide concentrated reagent has a specific gravity of 0.90, contains 29.0% NH_4OH, and is 15 N in concentration.

Add the amounts of conc NH_4OH indicated in Table 1.19 to about 400 mL of DDW with stirring in a hood. Dilute to 1 L.

1.12 How to Wash Laboratory Glassware

The washing of glassware frequently depends on the proposed use of the glassware. Often glassware can be color-coded so that segregation for the different uses and different concentration levels can be maintained. The following list of washing methods may be useful.

A. General Chemical Use Glassware

For most glassware used for routine analyses in a water quality laboratory, wash and scrub with $NaHCO_3$, rinse three times with tap water and three times with DDW.

B. Phosphorus and Nitrogen Nutrient Analyses Glassware

Wash with $NaHCO_3$ and rinse three times with tap water and two times with DDW. *Do not use detergent.* Soak overnight (or rinse) with 6 N HCl and rinse three or four times with fresh DDW.

C. Specific Chemical Analysis (TDS beakers, pipets)

Soak overnight in a chromic acid solution followed by $NaHCO_3$ wash, then rinse three times with tap water and three times with DDW. (The preparation of chromic acid solution requires extreme caution. Wear safety glasses, gloves, and protective clothing while making and using this solution. Place 75 mL of DDW in a 250 mL beaker. Add sodium dichromate [$Na_2Cr_2O_7$] while stirring until the solution is saturated [a few crystals will remain on the bottom]. Decant this solution into a 3 or 4 L beaker or Erlenmeyer flask set up on a magnetic stirrer. Slowly and carefully add 2.2 L [one 9 lb bottle] of concentrated sulfuric acid [H_2SO_4] to the saturated $Na_2Cr_2O_7$ solution. Stir carefully to dissolve.)

D. Biological Analyses Glassware

Wash with $NaHCO_3$ and rinse three times with tap water and three times with DDW. It may be necessary to autoclave the glassware. *Do not use detergent* or chromic acid soak.

E. Bacteriological Analyses Glassware

Wash with a phosphate-free laboratory detergent and hot water. Rinse at least three times with DDW. Autoclave for sterilization when necessary.

1.13 Cleaning of Spectrophotometer Cells

1. Clean the cells with a mild agent as soon as possible after each use.
2. Always start with DDW for aqueous solutions, or use any suitable organic solvent for organic materials.
3. Mild sulfonic detergents may be used if it is certain that they are true solutions and do not contain particulate matter.
4. For hard-to-remove deposits, use a solution of 50% 3 N HCl and 50% ethanol.
5. Remember that if a reagent is not of spectrograde purity it may leave a deposit on the cell window after evaporation.

NEVER blow the cell dry with air. It is better to speed evaporation of the solvent with the aid of a vacuum.

NEVER use any brush or instrument that might scratch the sides of the cell.

NEVER use alkali, abrasives, etching materials, or hot concentrated acids.

NEVER use ultrasonic devices to clean cells.

Table 1.1 Spiking Procedures.

Applicable Analyses:
Cr^{+6}, F (SPADNS), NH_3-N (low and high levels), NO_2^--N,NO_3^--N, TKN (Nesslerization), Total N (persulfate), PO_4^--P,SiO_2 and $SO_4^=$ (turbidimetric).

Procedure:
Prepare concentrated spike (CONSP) solution from the stock standard as outlined below. Add a known volume (usually 1.0 mL) of CONSP solution to the sample volume recommended in the analytical procedure.

Parameter	Concen-tration Level	Concentration Range of Sample to be Spiked	Concentration of Spike (CONSP), μg/mL	Stock Standard μg/mL	Dilution of Stock for Concentrated Spike (CONSP)
Cr^{+6}	high	50–100 μg/L	0.5	50.	2.0 mL → 100 mL
	interm.	10– 50 μg/L	1.0		4.0 mL → 100 mL
	low	<10 μg/L	0.1		1.0 mL → 500 mL
F^- (SPADNS)	high	1.0–1.4mg/L	5.0	100.	5.0 mL → 100 mL
	interm.	0.2–1.0mg/L	10.0		10.0 mL → 100 mL
	low	<0.2 mg/L	2.5		5.0 mL → 200 mL
NH_3-N (low)	high	300–400 μg/L	2.0	1000.	1.0 mL → 500 mL
	interm.	100–300 μg/L	5.0		1.0 mL → 200 mL
	low	<100 μg/L	1.0		1.0 mL → 1000 mL
NH_3-N (high)	high	30– 50 mg/L	250.	1000.	25.0 mL → 100 mL
	interm.	15– 30 mg/L	500.		50.0 mL → 100 mL
	low	<15 mg/L	100.		10.0 mL → 100 mL
NO_2^--N (manual or auto-mated)	high	75–100 mg/L	0.5	100.[a]	1.0 mL → 200 mL
	interm.	25– 75 mg/L	1.0		1.0 mL → 100 mL
	low	<25 mg/L	0.2		1.0 mL → 500 mL
NO_3^--N (auto-mated)	high	1.5–2.0 mg/L	10.0	100.	10.0 mL → 100 mL
	interm.	0.5–1.5 mg/L	25.0		25.0 mL → 100 mL
	low	<0.5 mg/L	5.0		5.0 mL → 100 mL
NO_3^--N (manual)	high	150–200 μg/L	2.0	1000.	1.0 mL → 500 mL
	interm.	50–150 μg/L	5.0		1.0 mL → 200 mL
	low	<50 μg/L	1.0		1.0 mL → 1000 mL
TKN[b] (Nessler-ization)	high	3.0–5.0 mg/L	25.0	1000.[b]	5.0 mL → 200 mL
	interm.	1.0–3.0 mg/L	50.0		5.0 mL → 100 mL
	low	<1.0 mg/L	10.0		1.0 mL → 100 mL
Total N[c] (Persul-fate)	high	7.0–10.0 mg/L	40.0	1000.[c]	4.0 mL → 100 mL
	interm.	2.0–7.0 mg/L	80.0		8.0 mL → 100 mL
	low	<2.0 mg/L	20.0		2.0 mL → 100 mL
PO_4^{-3}/P	high	500–700 μg/L	5.0	1000.	1.0 mL → 200 mL
	interm.	100–500 μg/L	10.0		1.0 mL → 100 mL
	low	<100 μg/L	2.0		1.0 mL → 500 mL

(*continued*)

Table 1.1, continued.

Parameter	Concen-tration Level	Concentration Range of Sample to be Spiked	Concentration of Spike (CONSP), μg/mL	Stock Standard μg/mL	Dilution of Stock for Concentrated Spike (CONSP)
SiO_2	high	30-40 mg/L	250.	1000.	25.0 mL → 100 mL
	interm.	10-30 mg/L	500.		50.0 mL → 100 mL
	low	< 10 mg/L	100.		10.0 mL → 100 mL
$SO_4^{=}$ (turbidi-metric)	high	20-30 mg/L	500.	1000.	50.0 mL → 100 mL
	interm.	10-20 mg/L	1000.		none; use stock
	low	< 10 mg/L	200.		20.0 mL → 100 mL

Notes: See Table 1.10 for precision and accuracy terms and calculations.

[a]Concentration ranges were determined from the automated method. Use the stock standard from the automated procedure (100 μg/mL).
[b]These calculations pertain to spiking after digestion and before Nesslerization only. To spike prior to digestion, see Table 1.2.
[c]Use the 1000 mg NH_3-N/L stock standard to prepare spiking solutions. Add 1.0 mL of CONSP solution before persulfate digestion.

Theoretical Concentration of Spiked Sample, TH-SP:

$$\text{TH-SP} = \frac{(\text{UNSP} \times S_1) + (\text{CONSP} \times S_2)}{S_3}$$

where

UNSP = Concentration of the unspiked sample in (μg or mg)/L; use $\bar{x}$ from precision data
CONSP = Concentration of spiking solution in (μg or mg)/mL prepare as outlined above

Note: Mass Units for UNSP and CONSP must be the same

S_1 = Sample volume in liters
S_2 = CONSP volume in mL; usually 1.0 mL
S_3 = Sample volume plus CONSP volume in liters

Experimental Concentration of Spike, EX-SP:

$$\text{EX-SP} = A\left(\frac{S_3}{S_1}\right)$$

where

A = Analyzed concentration of spiked sample in (μg or mg)/L
S_3 and S_1 are as defined above

$\frac{S_3}{S_1}$ = Factor to account for increased sample volume due to spike (CONSP)

Table 1.2 Spiking Procedures.

Applicable Analyses:

Ca Hardness, Total Hardness, Acidity, Alkalinity, Cl^-, CN^-, Total Phosphorus, COD (manual), TKN.

Procedure:

Using the table below, add a known volume (S_2) of concentrated spike solution (CONSP*) to the sample volume recommended in the analytical procedure.

*Unless otherwise indicated, the CONSP solution is the stock standard for that particular analysis.

Parameter	Concentration Level	Concentration Range of Sample to be Spiked	Stock Standard (CONSP), (μg or mg)/mL	mL (S_2) of CONSP* to Add to Each Sample
Ca Hardness and Total Hardness	high	125–200 mg/L	1.00 mg/mL (as $CaCO_3$)	2.0
	interm.	50–125 mg/L		5.0
	low	<50 mg/L		1.0
Acidity[a]	high	125–200 mg/L	~2.5 mg/mL[a] (as $CaCO_3$)	5.0
	interm.	50–125 mg/L		10.0
	low	<50 mg/L		2.0
Alkalinity[b]	high	350–500 mg/L	~2.50 mg/mL[b] (as $CaCO_3$)	2.0
	interm.	100–350 mg/L		4.0
	low	<100 mg/L		1.0
Cl^-	high	75–100 mg/L	0.50 mg/mL	5.0
	interm.	25-75 mg/L		10.0
	low	<25 mg/L		2.0
CN^-[c]	high	30–50 mg/L	1.00 mg/mL	5.0
	interm.	5–30 mg/L		10.0
	low	<5 mg/L		2.0
Total Phosphorus[d]	high	500–700 μg/L	1.00 μg/mL[d]	2.0
	interm.	100–500 μg/L		5.0
	low	<100 μg/L		1.0
COD[e] (manual, high level)	high	700–900 mg/L	1.00 mg/mL[e]	2.0
	interm.	200–700 mg/L		4.0
	low	<200 mg/L		1.0
COD[f] (manual, low-level)	high	75–100 mg/L	0.200 mg/mL[f]	2.0
	interm.	25–75 mg/L		5.0
	low	<25 mg/L		1.0
TKN[g] (manual)	high	7.0–10.0 mg/L	0.05 mg/mL[g]	2.0
	interm.	2.0–7.0 mg/L		4.0
	low	<2.0 mg/L		1.0
Surfactants[h] (MBAS)	high	5–10 mg/L	0.010 mg/mL[h]	2.0
	interm.	2–5 mg/L		4.0
	low	<2 mg/L		20.0

(Table continued on next page)

Table 1.2, continued

Notes: See Table 1.10 for precision and accuracy terms and calculations.

[a]Use the 0.05N KHP solution (~2.5 mg/L as $CaCO_3$) as the concentrated spiking solution (CONSP). Calculate the actual concentration of the CONSP solution as follows:

$$\text{mg } CaCO_3 \text{ /mL} = \text{CONSP} = 125. \times N$$

where:

N = The calculated normality of the ~0.02N NaOH; see acidity procedure, Standardization, step 1.a. and 1.b.

[b]Use 0.05 N Na_2CO_3 (~2500 mg/L as $CaCO_3$) as the spiking solution (CONSP). Calculate the actual concentration of the CONSP solution as follows:

$$\text{mg } CaCO_3 \text{ /mg} = \text{CONSP} = M \times 1.888$$

where

$$M = \text{g } Na_2CO_3 \text{ weighed into the 500 mL flask.}$$

[c]TH-SP formula applies if spike is added before digestion or after digestion prior to titration (sample volumes will be different). To spike before analysis by ion selective electrode, see Table 1.3.

[d]The Stock Standard is 1000 mg PO_4^{-3}-P/L. Dilute 1.0 mg of the Stock up to 1 L with DDW. Use this diluted stock (1.00 μg PO_4^{-3}-P/mL) as the CONSP solution. The spike is to be added prior to the digestion step. To spike after digestion, follow the procedure for PO_4^{-}-P, Table 1.1.

[e]TH-SP formula applies if the spike is added before or after digestion. Use the Stock KHP Standard from the low-level ampule COD method (1000 mg COD/L) as the CONSP solution.

[f]TH-SP formula applies if the spike is added before or after digestion. Using the Stock KHP Standard (1000 mg COD/L) from the low-level ampule COD method, dilute 20.0 mL up to 100 mL with DDW. This solution is 0.200 mg COD/mL and is to be used as the CONSP solution.

[g]TH-SP formula applies only if the spike is added before digestion. (To spike before Nesslerization, see Table 1.1). Using the Stock NH_4Cl Solution (1000 mg NH_3-N/L), dilute 5.0 mL up to 100 mL with DDW. This solution is 0.05 mg NH_3-N/L and is to be used as the CONSP solution.

[h]Use the Standard LAS Solution (10 mg LAS/L) as the concentrated spiking solution (CONSP).

Theoretical Concentration of Spiked Sample, TH-SP:

$$\text{TH-SP} = \text{UNSP} + \left(\frac{\text{CONSP} \times S_2}{S_1}\right)$$

where

UNSP = Concentration of the unspiked sample in (μg or mg)/L; use $\bar{x}$ from precision data
CONSP = Concentration of spiking solution in (μg or mg)/mL as indicated above

Note: Mass units for UNSP and CONSP must be the same
S_1 = Sample volume in liters
S_2 = mL of CONSP solution added

Table 1.3 Spiking Procedures.

Applicable Analyses:

CN^- (ion-selective electrode), F^- (ion selective electrode).

Procedure:

Using the table below, add a known volume (S_2) of concentrated spike solution (CONSP) to the sample volume recommended in the analytical procedure. The CONSP solution is the stock standard for that particular analysis.

Parameters	Concentration Level	Concentration Range of Sample to be Spiked (mg/L)	Concn. of Spike (CONSP), mg/mL	mL (S_2) of CONSP to add to each sample
CN^- [a] (electrode)	high	30–50	1.00	10.0
	interm.	5–30		20.0
	low	<5		5.0
F^- (electrode)	high	1.0–2.0	0.01	5.0
	interm.	0.2–1.0		10.0
	low	<0.2		2.0

Notes: See Table 1.10 for precision and accuracy terms and calculations.

[a] TH-SP formula applies only if spike is added after digestion. (To spike before digestion, see Table 1.2).

Theoretical Concentration of Spiked Sample, TH-SP:

$$\text{TH-SP} = \text{UNSP} + (\text{CONSP} \times S_2)$$

where

UNSP = Concentration of the unspiked sample in (μg or mg)/L; use $\bar{x}$ from precision data
CONSP = Concentration of spiking solution in (μg or mg)/mL as indicated above.

Note: Mass units for UNSP and CONSP must be the same.

S_2 = mL of CONSP solution added

Table 1.4 Spiking Procedures.

Applicable Analysis:
TS, TDS (Residue II), $SO_4^{=}$ (gravimetric), Oil and Grease.

Procedure:
Weigh a quantity of spiking compounds, as indicated in the table below, to the nearest 0.1 mg. Add to the sample volume recommended in the analytical procedure.

Parameter	Concentration Level	Concentration Range of Sample to be Spiked (mg/L)	Spiking Compound	Approx. Mass of Spiking Compound to Add to Each Sample (mg)
TDS	high	>5000	KCl[a]	50
	interm.	500–5000	Na_2SO_4, etc.	100
	low	<500		20
$SO_4^{=}$ (gravimetric)	high	>2000	Na_2SO_4[b]	50
	interm.	500–2000		190
	low	<500		20
Oil and Grease	high	500–1000	Vegetable[c]	150
	interm.	50–500	Oil, Fuel	300
	low	<50	Oil, etc.	50

Notes: See Table 1.10 for precision and accuracy terms and calculations.

[a]Choose a spiking compound of high solubility in water. Dry in a 103°C oven and cool in a desiccator before weighing. Avoid carbonate salts, since a portion of the weighed amount may be lost during the drying process.
[b]Dry the spiking compound in a 103°C oven and cool in a desiccator before weighing.
[c]Choose an oil for which high recovery (90–100%) is expected. Consult *Standard Methods* (1989, pp. 5–41 through 5–46).

Theoretical Concentration of Spiked Sample, TH-SP:

$$\text{TH-SP} = \text{UNSP} + \left(\frac{M}{S_1}\right)$$

where

UNSP = concentration of the unspiked sample in mg/L; use $\bar{x}$ from precision data
M = mass of spiking compound added to sample in mg
S_1 = sample volume in liters

Table 1.5 Spiking Procedures.

Applicable Analysis:
Specific Conductance.
Procedure:
Dry the KCl spiking compound at 103° and cool in a desiccator. Weigh a quantity of KCl, as indicated below, to the nearest 0.1 mg. Add the KCl to 100.0 mL of sample (or a known sample volume which is adequate to cover the conductivity probe).

Concentration Level	Concentration Range of Sample to be Spiked (μmhos/cm)	Approximate Mass of KCl to Add to Each Sample (mg)
high	>10,000	250
interm.	3000–10,000	100
low	<3,000	50

Notes: See Table 1.10 for precision and accuracy terms and calculations.

Theoretical Concentration of Spiked Sample, TH-SP:

$$\text{TH-SP} = \text{UNSP} + \left(\frac{\text{M} \times 1.895}{\text{S}_1}\right)$$

where

UNSP = concentration of unspiked sample in μmhos/cm at 25°C; use $\bar{x}$ from precision data
M = mass of KCl added to sample in mg
S_1 = sample volume in liters
1.895 = factor to convert weight of KCl to μmhos/cm.

Table 1.6 Spiking Procedures.

Applicable Analysis:
Boron
Procedure:
Prepare concentrated spike (CONSP) solutions as outlined below. Add 1.0 mL of CONSP solution to 1.0 mL of sample and carry through the analytical procedure.

Concentration Level	Concentration Range of Sample to be Spiked (μg/mL)	Concentration of Spike (CONSP), μg/mL	Stock Standard, μg/mL	Dilution of Stock for Concentrated Spike (CONSP)
high	2000–3000	1.0	100.	1.0 mL → 100 mL
interm.	500–2000	2.0		2.0 mL → 100 mL
low	<500	0.5		1.0 mL → 200 mL

Notes: See Table 1.10 for precision and accuracy terms and calculations.

Theoretical Concentration of Spiked Sample, TH-SP:

$$\text{TH-SP} = \frac{\text{UNSP} + (\text{CONSP} \times 1000)}{2}$$

where
UNSP = concentration of the unspiked sample in μg/L; use $\bar{x}$ from precision data
CONSP = concentration of spiking solution in μg/mL as prepared above.

Table 1.7 Spiking Procedures.

Applicable Analysis:
COD (Ampule; High and Low-Level).
Procedure:
Prepare concentrated spike (CONSP) solutions from the high-level stock standard as outlined below. Decrease the sample volume used for the precision data (S_1) by 1.0 mL. To this sample volume add 1.0 mL of CONSP solution and carry through the analytical procedure.

Parameter	Concentration Level	Concentration Range of Sample to be Spiked (mg/L)	Concentration of Spike (CONSP), mg/mL	Stock Standard mg/mL	Dilution Stock for Concentrated Spike (CONSP)
COD	high	700–990	1.0	10.0	10.0 mL → 100 mL
(ampule,	interm.	200–700	2.0		20.0 mL → 100 mL
high-level)	low	<200	0.5		5.0 mL → 100 mL
COD	high	75–100	0.10	10.0	1.0 mL → 100 mL
(ampule,	interm.	25–75	0.20		2.0 mL → 100 mL
low-level)	low	<25	0.05		1.0 mL → 200 mL

Notes: See Table 1.10 for precision and accuracy terms and calculations.

Theoretical Concentration of Spiked Sample, TH-SP:

$$\text{TH-SP} = \frac{\text{UNSP}(S_1 - 0.001) + \text{CONSP}}{S_1}$$

where

UNSP = concentration of the unspiked sample in mg/L; use $\bar{x}$ from precision data
CONSP = concentration of spiking solution in mg/mL as prepared above
S_1 = sample volume in liters (0.002, 0.0025, or 0.003L as indicated in procedure).

Table 1.8 Spiking Procedures.

Applicable Analyses:
Residue (I and II), Turbidity, Chlorine (residual free), BOD, Chlorophyll *a* (fluorometric and spectrophotometric).
Procedure:
Mix equal volumes of the sample and an internal quality control. Carry the spiked sample through the analytical procedure.

Notes: See Table 1.10 for precision and accuracy terms and calculations.

Theoretical Concentration of Spiked Sample, TH-SP:

$$\text{TH-SP} = \frac{\text{UNSP} + \text{QC}}{2}$$

where

UNSP = concentration of the unspiked sample in (μg or mg)/L (or NTU); use $\bar{x}$ from precision data
QC = "true" value of the internal quality control in (μg or mg)/L (or NTU)

Mass units for UNSP and QC must be the same.

Table 1.9 Example of Precision Data and Statement.

Quality Control Data; Precision

Parameter: AMMONIA, NH3-N, LOW LEVEL Date: 9-3-88

Method: INDOPHENOL Analyst: B.C.

Reference: WATER + WASTEWATER EXAMINATION MANUAL

concentration (μg, mg/L)

n	Sample # 5709-88	Sample # 5701-88	Sample # 5713-88	Sample # 5699-88
1	389.	15.	247.	131.
2	390.	12.	252.	135.
3	387.	12.	255.	125.
4	383.	14.	245.	128.
5	390.	10.	250.	133.
6	397.	13.	255.	133.
7	380.	15.	243.	134.
8				
9				
10				
$\bar{x}$	388.	13.	250.	131.
s	±5.5	±1.8	±4.8	±3.6
RSD	1.4%	14%	1.9%	2.7%

Notes: Table 1.10 outlines precision terms and calculations needed above. *Precision Statement:* Using water samples at concentrations of 13. (μg) mg/L and 388. (μg) mg/L NH_3-N, the standard deviations were ±1.8 and ±5.5 units, respectively. The relative standard deviations were 14 % and 1.4 %, respectively.

Table 1.10 Precision and Accuracy Terms and Calculations.

Note: Use Tables 1.1 through 1.8 to calculate the theoretical concentration of the spiked sample, TH-SP in (μg or mg)/L or other appropriate units. Use the terms and equations below for collecting precision and accuracy data:

TH-SP = theoretical concentration of spiked sample

UNSP = concentration of the unspiked sample in (μg or mg)/L, NTU, or μmhos/cm; use the mean, $\bar{x}$, from precision data

CONSP = concentration of the spiking solution in (μg or mg)/mL; see Tables 1.1, 1.2, 1.3, 1.6, and 1.7

M = mass of spiking compound in mg; see Tables 1.4 and 1.5

QC = "true" value of the internal quality control in (μg or mg)/L, or NTU. See Table 1.8

S_1 = sample volume in liters

S_2 = CONSP volume added in mL

S_3 = sample volume plus CONSP volume in liters (for Table 1.1 only)

EX-SP = experimentally determined concentration of the spiked sample in (μg or mg)/L or NTU (exception, see Table 1.1)

%D = percent deviation of the experimentally determined spiked sample from the theoretical concentration

$$= \left[\frac{(\text{EX–SP}) - (\text{TH–SP})}{\text{TH-SP}}\right] \times 100$$

Note: Retain sign as positive or negative

%R = percent recovery of the spike

= 100 + %D

$\overline{\%R}$ = mean percent recovery

$$= \frac{\Sigma\ \%R_i}{n}$$

where

n = the number of replicates

$\%R_i$ = the %R for each n

$\bar{x}$ = mean value for precision data

$$= \frac{\Sigma x_i}{n}$$

where

n = the number replicates

x_i = the x value for each n

s = standard deviation of the percent recovery or precision data

$$= \sqrt{\frac{\Sigma\ (\%R_i - \overline{\%R})^2}{n-1}} \quad \text{or} \quad \sqrt{\frac{\Sigma\ (x_i - \bar{x})^2}{n-1}}$$

RSD = relative standard deviation of the percent recoveries or precision data (%)

$$= \frac{s}{\overline{\%R}} \times 100 \quad \text{or} \quad \frac{s}{\bar{x}} \times 100$$

Table 1.11 Replicate Sample Analyses for Ammonia, NH_3-N.

Sample	Replicate	NH_3-N, μg/L	Mean, $\bar{x}$, μg/L	Percent Deviation[a] from the Mean
6309–88	1	202.	228	11.6
	2	255.		
6319–88	1	11.	16.	31.2
	2	21.		
6329–88	1	425	431.	1.4
	2	437.		
6339–88	1	98.	102.	3.4
	2	105.		

Note:

$\bar{x}$ = mean value

$$= \frac{\Sigma x_i}{n}$$

where

n = the number of replicates

x_i = the x value for each n

[a]Percent deviation from the mean $= \frac{x_i - \bar{x}}{\bar{x}} \times 100.$

Table 1.12 Example of Accuracy Data and Statement.

Quality Control Data; Accuracy

Parameter: Ammonia, NH_3-N, Low Level Method: Indophenol

Reference: Water & Wastewater Exam. Manual Analyst: Bobby Chemist Date: 9-3-88

Sample #: 5709-88				Sample #: 5701-88			Sample #: 5713-88			Sample #: 5699-88		
UNSP = 388. (µg, mg)/L				UNSP = 13. (µg, mg)/L			UNSP = 250. (µg, mg)/L			UNSP = 131. (µg, mg)/L		
CONSP = 2. (µg, mg)/L				CONSP = 1. (µg, mg)/L			CONSP = 5. (µg, mg)/L			CONSP = 5. (µg, mg)/L		
TH-SP = 420. (µg, mg)/L				TH-SP = 32. (µg, mg)/L			TH-SP = 343. (µg, mg)/L			TH-SP = 226 (µg, mg)/L		
n	EX-SP	%D	%R	EX-SP	%D	%R	EX-SP	%D	%R	EX-SP	%D	%R
1	409.	-2.6	97.4	28.	-12.5	87.5	338.	-1.5	98.5	219.	-3.0	97.0
2	430.	+2.4	102.4	29.	-9.4	90.6	344.	+0.3	100.3	222.	-1.8	98.2
3	412.	-1.9	98.1	37.	+15.6	115.6	338.	-1.5	98.5	233.	+3.1	103.1
4	417	-0.7	99.3	30.	-6.2	93.8	333.	-2.9	97.1	233.	+3.1	103.1
5	409.	-2.6	97.4	35.	+9.4	109.4	326.	-5.0	95.0	234.	+3.5	103.5
6	404.	-3.8	96.2	37.	+15.6	115.6	339.	-1.2	98.8	227.	+0.4	100.4
7	407.	-3.1	96.9	36.	+12.5	112.5	331.	-3.5	96.5	235.	+4.0	104.0
8												
9												
10												
$\overline{\%R}$			98.2			103.6			97.8			101.3
s			±2.1			±12.4			±1.7			±2.8
RSD			2.1%			12.0%			1.7%			2.8%

Notes: This form is applicable to spiking procedures for analyses listed in Tables 1.1, 1.2, 1.3, 1.6, and 1.7. Calculate TH-SP by consulting these tables. Table 1.10 outlines other accuracy terms and calculations needed above. S_1 = 0.050 ℓ; S_2 = 1.0 mℓ; S_3 = 0.051 ℓ.

Accuracy Statement: Using water samples at concentrations of 13. (µg) mg/L and 388 (µg) mg/L, NH_3-N, the recoveries were 103.6 % and 97.2 %, respectively. The standard deviations for percent recoveries were ± 12.4 and ± 2.1 units, respectively. The relative standard deviations were 12.0 % and 2.1 %, respectively.

Table 1.13 Example of Accuracy Data and Statement.

Quality Control Data; Accuracy

Parameter: Sulfate Method: Gravimetric
Reference: Water & Wastewater Exam. Manual Analyst: Bobby Chemist Date: 9-17-88
Spiking Compound: Na_2SO_4 S_1: 0.150l

	Sample #: 5833-88 UNSP = 2230. (μg, mg)/L Approx. M 50. mg QC = ____ (μg, mg)/L				Sample #: 5832-88 UNSP = 429. (μg, mg)/L Approx. M 20. mg QC = ____ (μg, mg)/L				Sample #: 5840-81 UNSP = 1480. (μg, mg)/L Approx. M 100. mg QC = ____ (μg, mg)/L				Sample #: 5838-88 UNSP = 964. (μg, mg)/L Approx. M 100. mg QC = ____ (μg, mg)/L			
n	M	TH-SP	EX-SP	%R	M	TH-SP	EX-SP	%R	M	TH-SP	EX-SP	%R	M	TH-SP	EX-SP	%R
1	49.0	2560.	2430.	94.9	25.1	596.	610.	102.3	101.2	2150.	2140.	99.5	107.5	1680.	1650.	98.2
2	50.3	2570.	2490.	96.9	20.3	564.	552.	97.9	100.7	2150.	2220.	103.3	100.0	1630.	1650.	101.2
3	49.1	2560.	2520.	98.4	19.5	559.	541.	96.8	99.6	2140.	2180.	101.9	102.5	1650.	1720.	104.2
4	49.7	2560.	2590.	101.2	18.6	553.	532.	96.2	97.5	2130.	2090.	98.1	96.4	1610.	1650.	102.5
5	52.0	2580.	2470.	95.7	21.7	574.	548.	95.5	95.0	2110.	2210.	104.7	101.9	1640.	1640.	100.0
6	51.5	2570.	2480.	96.5	21.8	574.	583.	101.6	97.8	2130.	2160.	101.4	84.5	1590.	1650.	103.8
7	47.2	2540.	2680.	105.6	20.5	566.	556.	98.2	102.3	2160.	2140.	99.1	106.3	1670.	1760.	105.4
8																
9																
10																
$\overline{\%R}$				98.5				98.4				101.1				102.2
s				±3.8				±2.6				±2.4				±2.5
RSD				3.9%				2.6%				2.4%				2.5%

Notes: This form is applicable to spiking procedures for analyses listed in Tables 1.4, 1.5, and 1.8 Calculate TH-SP by consulting these tables. When using a spiking compound, TH-SP values will be different for each replicate spiked sample. Table 1.10 outlines other accuracy terms and calculations needed above. *Accuracy Statement:* Using water samples at concentrations of 429. μg, mg/L and 2230. μg, mg/L $SO_4^=$, the recoveries were 98.4 % and 98.5 %, respectively. The standard deviations for percent recoveries were ± 2.6 and ± 3.8 units, respectively. The relative standard deviations were 2.6 % and 3.9 %, respectively.

Table 1.14 Spiked Sample Analysis for Ammonia, NH_3-N.

Sample	TH-SP (μg/L)	EX-SP (μg/L)	%R	%Deviation[a]
6309–88	381.	353.	90.3	-8.0
6319–88	282.	276.	97.9	-3.4
6329–88	331.	309.	93.4	-4.5
6339–88	78.	77.	98.4	-4.7

Notes: See Table 1.10 for precision and accuracy terms and calculations.

[a]Percent deviation of %R (determined daily) from the $\overline{\%R}$ (accuracy form, Tables 1.12 or 1.13) in the sample concentration range of the sample spiked:

$$\left(\frac{\%R - \overline{\%R}}{\overline{\%R}}\right) \times 100$$

Table 1.15 Recommendations for Preservation of Samples.

Measurement	Vol. Req. (mL)	Container[a]	Preservative	Recommended Holding Time[b,c,d]
Physical Properties				
Conductance	100	P,G	Cool, 4° C	28 days[e]
pH	25	P,G	Determine on site	2 hours
Residue				
Filterable	100	P,G	Cool, 4° C	7 days
Non-Filterable	100	P,G	Cool, 4° C	7 days
Total	100	P,G	Cool, 4° C	7 days
Volatile	100	P,G	Cool, 4° C	7 days
Settleable Matter	1000	P,G	None required	48 hrs
Temperature	100	P,G	Determine on site	No holding
Turbidity	100	P,G	Cool, 4° C	48 hours
Metals				
Dissolved	200	P,G	Filter on site HNO_3 to pH<2	6 months[f]
Suspended	200		Filter on site	6 months
Total	100	P,G	HNO_3 to pH<2	6 months[f]
Hardness	100	P,G	Cool, 4° C HNO_3 to pH<2	6 months[f]
Mercury				
Dissolved	100	P,G	Filter on site HNO_3 to pH<2	28 days
Total	100	P,G	HNO_3 to pH<2	28 days

Table 1.15, continued.

Measurement	Vol. Req. (mL)	Container[a]	Preservative	Recommended Holding Time[b,c,d]
Inorganics, Non-Metallics				
Acidity	100	P,G	Cool, 4° C	14 days
Alkalinity	100	P,G	Cool, 4° C	14 days
Chloride	100	P,G	None required	28 days
Chlorine	200	P,G	Determine on site	No holding; analyze immediately
Cyanides	500	P,G	Cool, 4° C NaOH to pH 12	14 days
Fluoride	300	P	None required	28 days
Nitrogen, Ammonia	400	P,G	Cool, 4° C H_2SO_4 to pH<2	28 days
Kjeldahl, Total	500	P,G	Cool, 4° C H_2SO_4 to pH<2	28 days
Nitrate plus Nitrite	100	P,G	Cool, 4° C H_2SO_4 to pH<2	28 days
Nitrate	100	P,G	Cool, 4° C or Chloroform, 4° C	48 hours
Nitrite	50	P,G	Cool, 4° C Chloroform, 4° C	48 hours
Dissolved Oxygen				
Probe	300	G only, BOD bottle	Determine on site	No holding; analyze immediately
Winkler	300	G only, BOD bottle	Fix on site	4–8 hours, after acidified
Phosphorus Orthophosphorus Dissolved	50	P,G	Filter on site Cool, 4° C	48 hours
Hydrolyzable	50	P,G	Cool, 4° C H_2SO_4 to pH<2	48 hours
Total Phosphorus				
Total	50	P,G	Cool, 4° C H_2SO_4 to pH 2	24 hours
Dissolved	50	P,G	Filter on site, Cool, 4° C H_2SO_4 to pH<2	24 hours

Table 1.15, continued.

Measurement	Vol. Req. (mL)	Container[a]	Preservative	Recommended Holding Time[b,c,d]
Silica	50	P only	Cool, 4° C	28 days
Sulfate	50	P,G	Cool, 4° C	28 days
Sulfite	50	P,G	Determine on site	No holding; analyze immediately
Organics				
BOD	1000	P,G	Cool, 4° C	48 hours
COD	50	P,G	H_2SO_4 to pH<2	28 days
Oil and Grease	1000	G only	Cool, 4° C H_2SO_4 to pH<2	28 days
Organic carbon	25	S	Cool, 4° C H_2SO_4 to pH<2	28 days
MBAS	250	P,G	Cool, 4° C	48 hours

[a]Plastic (p) or Glass (G). For most metals, polyethylene with a polypropylene cap (no liner) is preferred.
[b]It should be pointed out that holding times listed above are recommended for properly preserved samples based on currently available data. It is recognized that for some sample types, extension of these times may be possible while for other types, these times may be too long. Where shipping regulations prevent the use of the proper preservation technique or the holding time is exceeded, such as the case of a 24-hour composite, the final reported data for these samples should indicate the specific variance.
[c]*Standard Methods for the Examination of Water and Wastewater*, 1989, 17th ed. American Public Health Association, Washington, DC.
[d]EPA. 1983. *Methods for Chemical Analysis of Water and Wastes.* U.S. Environmental Protection Agency, EPA-690/4–79–020, Cincinnati, OH.
[e]If the sample is stabilized by cooling, it should be warmed to 25° C for reading, or temperature correction made and results reported at 25° C.
[f]Where HNO_3 cannot be used because of shipping restrictions, the sample may be initially preserved by icing and immediately shipped to the laboratory. Upon receipt in the laboratory, the sample must be acidified to a pH< 2 with HNO_3 (normally 3 mL 1 + 1 HNO_3/L is sufficient). At the time of analysis, the sample container should be thoroughly rinsed with 1 + 1 HNO_3 and the washings added to the sample (volume correction may be required).

Table 1.16 Suggestions for Filtering.

Test	Unfiltered	Filter/Type of Filter
Physical Properties		
Specific Conductance	√	
Nonfilterable Residue	√	
Total Residue	√	
Total Filterable Residue (TDS)		√ GF/C[a]
Turbidity	√	
Settleable Matter	√	
Metals		
Total Metals (Acid Digestion)	√	
Dissolved Metals		√ Membrane; filter[b] *before* acid preservation.
Inorganics, Non-Metals		
Acidity	√	
Alkalinity	√	
Chloride	√	√ Membrane[b]
Chlorine	√	
Cyanide	√	
Fluoride	√	√ Membrane[b]
Nitrogen		
Ammonia-Distillation	√	
Ammonia-Low or High Level	√	√ Membrane[b]
Nitrate, Nitrite		√ Membrane[b]
TKN, Total N	√	
Oxygen, Dissolved	√	
pH	√	
Phosphorus		
Orthophosphorus		√ Membrane[b]
Total Phosphorus	√	
Silica	√	√ Membrane[b]
Sulfate	√	√ Membrane[b]
Sulfite	√	
Organics		
BOD	√	
COD	√	
Oil and Grease	√	
Surfactants	√	
Total Organic Carbon (Persulfate)	√	
Biological		
Chlorophyll *a*	√	
Plankton Pigments	√	
Total Coliform	√	
Fecal Coliform	√	
Heterotrophic (Standard) Plate Count	√	

[a]Glass Fiber Filter: Whatman GF/C or equivalent.
[b]Membrane Filter: Millipore HA (0.45μm) or equivalent.

Table 1.17 Preparation of Acid Solutions.

	Hydrochloric Acid (HCl)	Sulfuric Acid (H_2SO_4)	Nitric Acid (HNO_3)
Specific Gravity (20/4°C) of American Chemical Society (ACS) Grade Concentrated Acid	1.174–1.189	1.834–1.836	1.490–1.418
Percent of active ingredient in concentrated acid	36–37	96–98	69–70
Normality of concentrated acid	11–12	36	15–16
Volume (mL) of concentrated acid to prepare 1 L of:			
18N solution	. . .	500	. . .
6N solution	500	167	380
1N solution	83	28	64
0.1N solution	8.3	2.8	6.4

Source: Standard Methods for the Examination of Water and Wastewater (17th edition). Copyright 1989 by APHA, AWWA, and WPCF. Reprinted with permission.
Notes: For a more accurate 0.1 N solution, dilute 17 mL of 6 N acid solution to 1 L. To prepare a 0.02 N acid solution, dilute 20 mL of 1 N reagent to 1 L.

Table 1.18 Preparation of Hydroxide Solutions.

Normality of Hydroxide Solution	Weight of Chemical Required to Prepare 1 L of Solution	
	Sodium Hydroxide (NaOH)(g)	Potassium Hydroxide (KOH)(g)
6	240	357
2	30	112
1	40	56
0.1	4	6

Source: Standard Methods for the Examination of Water and Wastewater (17th edition). Copyright 1989 by APHA, AWWA, and WPCF. Reprinted with permission.
Note: Never store NaOH or KOH in glass containers with ground glass stoppers. The glass will become etched and the stopper frozen in position. Reagent grade NaOH and KOH are not of primary standard grade. If these solutions are to be used as titrants where the exact normality is required, they must be standardized.

Table 1.19 Preparation of Ammonium Hydroxide Solutions.

Normality of NH_4OH Solutions	mL of Concentrated NH_4OH Required to Prepare 1 L of Solution
5	333
3	200
1	67
0.2	13

Methods for Determining Physical Properties

2. Methods for Determining Physical Properties

2.1 Specific Conductance

A. General

1. References: See *Standard Methods* (1989, pp. 2–57 through 2–61) and EPA (1983, Method 120.1–1).
2. Outline of Method:
   Specific conductance measures a water's capacity to carry an electric current. This property is related to the total concentration of the ionized substances in the water and the temperature at which the measurement is made. Specific conductance is used to monitor the quality of deionized water. The amount of dissolved ionic matter in a sample can be estimated by multiplying the conductivity by some empirical factor. Specific conductance should be reported as μmhos/cm at 25°C.

B. Special Reagents

Standard Potassium Chloride, 0.0100 M: Dissolve 745.6 mg anhydrous KCl in freshly boiled DDW and dilute to 1 L. At 25°C this solution has a specific conductance of 1413 μmhos/cm. Store in a glass-stoppered Pyrex bottle.

C. Standardization

1. Measure the temperature and the conductivity of standard solution, K_{std}, daily.

D. Procedure

1. Assemble conductivity meter and rinse the electrode thoroughly with DDW.
2. Measure the temperature and the conductivity of samples, K_{sam}.

E. Calculations

1. If temperatures of the samples and the standard are the same,

$$\text{Specific Conductance} = \frac{1413.}{K_{std}} K_{sam} \mu\text{mhos/cm}$$

2. If temperatures are different, correct all readings, including K_{std}, to 25°C (see Figure 2.1 and Table 2.1), and then calculate specific conductance by the above formula.

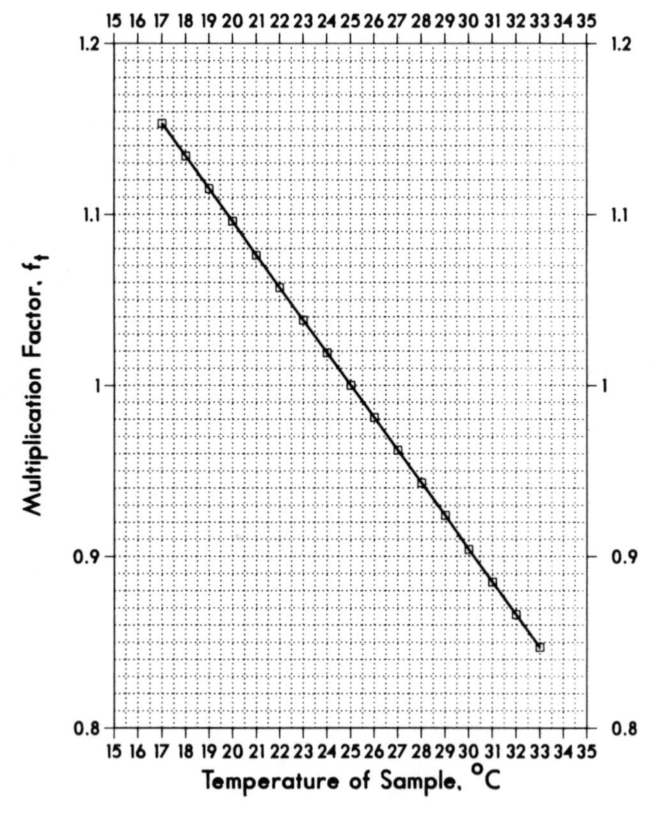

Figure 2.1 Relationship of multiplication factor and temperature of sample to correct conductivity measurements to 25° C.

Table 2.1 Factors for Converting Specific Conductance of Water to Equivalent Values at 25°C.

°C	f_t
17	1.153
18	1.334
19	1.115
20	1.096
21	1.076
22	1.057
23	1.038
24	1.019
25	1.000
26	0.981
27	0.962
28	0.943
29	0.924
30	0.904
31	0.885
32	0.866
33	0.847

Notes: $K_{25} = K_t \times f_t$. The greater the deviation of the temperature from 25°C, the greater the uncertainty in applying the temperature correction. K_{25} = conductivity of solution at 25°C; K_t = conductivity of solution at t°C; f_t = multiplication factor at t°C.

2.2 Residue I: Nonfilterable Residue

A. Total Suspended Solids Dried at 103–105°C
B. Volatile Suspended Solids Ashed at 550°C

A. General

1. Reference: See *Standard Methods* (1989, pp. 2-75 through 2-77) and EPA (1983, Methods 160.2 and 160.4).
2. Outline of Method:
   Total Suspended Solids—Total suspended solids are the residues retained on a glass fiber filter after filtration. The filter and solids are dried at 103 to 105°C to a constant weight.
   Volatile Suspended Solids—The volatile components of the nonfilterable residue are determined by igniting the filter and solids at 550°C after determining the suspended solids. The loss of weight after ignition is the volatile suspended solids.

B. Special Apparatus and Equipment

1. *Filter Disks Glass-Fiber Filters:* Use Whatman GF/C or equivalent 4.25 cm disc size.
2. Use an *Analytical Balance.*
3. Use a *Drying Oven and Muffle Furnace.*
4. Use *Flat-bladed Forceps without Serrated Tips.*
5. *Filtration Setup:* This consists of a side arm flask, vacuum source, funnel, and filter holder.

C. Standardization

Run two blanks following the procedure below using DDW instead of a sample. The blanks help to correct for errors that result because filters are hygroscopic and may pick up or lose water from the air.

D. Procedure

Total Suspended Solids:

1. Filter Preparation: Place the filter on a vacuum filtration setup and filter approximately 100 mL of DDW. Dry filter to constant weight in the oven at 103°C. For volatile solids determination also, dry to constant weight in a muffle furnace for 15 to 30 min at 550°C. Cool the filter to room temperature in a desiccator after removal from the oven. Weigh to the nearest 0.1 mg using an analytical balance, making sure the pan is clean. Place the filter in an aluminum weighing boat (the boat will not be part of the initial weight) and record the sample (or filter) number on the boat.

   Note: Use smooth forceps at all times to handle filters. Pick them up carefully only near the edges.

2. Carefully place preweighed filter on a clean filtration setup.
3. Filter as much sample as will easily pass through the filter. Rinse down the funnel with a small amount of DDW. Never add so much sample that the filter becomes clogged and some sample has to be discarded from the funnel.
4. Record the filter number (or sample number), filter weight in grams, and sample volume in liters.
5. Place the filters in the 103°C oven using aluminum weighing boats. Dry to constant weight. (This process requires a minimum of 1 hr drying time. The usual process is to leave the filters in the oven overnight.)
6. Cool the filter in a desiccator and weigh. Record the weight. Proceed to volatile suspended solids determination if appropriate.

Volatile Suspended Solids:

1. After the final weighing for suspended solids, place the filter in a suitable container and ash to constant weight at 550°C in a muffle furnace. (This process usually requires 30 min.) Always allow the muffle furnace to come to temperature before igniting the filters.

   Note: The filter must be pre-ashed before the initial weighing in volatile suspended solids determination (see 2.2 D, step 1).

2. Allow the filter to cool to room temperature in a desiccator; weigh the filter and record the weight.

E. Calculation

1. Total Suspended Solids

$$\text{mg Total Suspended Solids/L} = \frac{(A - B - C) \times 1000}{\text{sample volume, L}}$$

where

A = weight of filter and solids, grams
B = weight of filter, grams
C = blank correction, grams
 = (final weight of blank filter) - (initial weight of blank filter)

2. Volatile Suspended Solids

$$\text{mg Volatile Suspended Solids/L} = \frac{(A - B - C) \times 1000}{\text{sample volume, L}}$$

where

A = weight of filter and solids before ignition, grams
B = weight of filter and solids after ignition, grams
C = blank correction, grams
 = (weight of blank filter before ignition) - (weight of blank filter after ignition)

2.3 Residue II: Total and Total Filterable

A. Total Solids
B. Total Dissolved Solids

A. General

1. References: See *Standard Methods* (1989, pp. 2-71 through 2-75) and EPA (1983, Methods 160.1 and 160.3).
2. Outline of Method:
   An unfiltered or filtered sample is evaporated to dryness and the material remaining is total or total dissolved solids. Total dissolved solids is defined as the material that passes through a standard glass fiber filter and remains after evaporation.

B. Special Apparatus and Equipment

1. *Filter Disks Glass-Fiber Filters:* Use Whatman GF/C or equivalent 4.25 cm disc.
2. Use *Drying Ovens*.
3. Use an *Analytical Balance*.

C. Standardization

Run two blanks following the procedure below using DDW instead of a sample.

D. Procedure

For total solids use unfiltered sample. For total dissolved solids filter the sample using a GF/C filter.

1. Any nonhomogeneous materials should be removed from the sample. Disperse visible floating oil and grease with a blender before analysis.
2. Choose an aliquot of sample that contains a residue of at least 25 mg.

3. Clean a 150 or 250 mL beaker (Pyrex) in chromic acid cleaning solution (see 1.12 C) and rinse well with DDW. Dry to constant weight in an oven overnight. Use a 103°C oven for total solids and a 180°C oven for total dissolved solids.
4. Cool in a desiccator and weigh to determine initial weight.
5. Measure out an aliquot of total or filtered sample into a beaker; do not handle the beaker directly.
6. Place in the oven at 103°C and evaporate to dryness. (For total solids, dry to constant weight).
7. For total dissolved solids, dry to constant weight in a 180°C oven. (This process usually requires 1 hr.)
8. Cool in a desiccator and weigh. Record final weight.

E. Calculation

For Total Solids (unfiltered sample) or Total Dissolved Solids (filtered sample)

$$\text{mg Solids/L} = \frac{(A - B - C) \times 1000}{\text{volume of sample, L}}$$

where

A = final weight = beaker + sample after evaporation
B = initial weight = beaker weight
C = blank correction
= (final weight of blank) - (initial weight of blank)

2.4 Settleable Matter
By Weight

A. General

1. Reference: See *Standard Methods* (1989, p. 2–78) and EPA (1983, Method 160.5).
2. Outline of Method:
   This is a measure of the "settling qualities" of suspended solids and measures the portion of material that settles within 1 hr. It also includes the material that floats to the surface.

B. Special Reagents: None

C. Standardization: None

D. Procedure

1. Measure suspended solids (see 2.2) in the total sample.
2. Place the well-mixed sample in a cylinder having ≥9 cm diameter, ≥20 cm height, and volume ≥1300 mL.
3. After 1 hr, siphon ~250 mL of sample from a point halfway between the floatable materials and the settled materials. Do not disturb the surface or bottom material when sampling.
4. Measure suspended solids on this material (the nonsettling matter).

E. Calculation

mg/L Settleable Matter = (mg/L suspended solids) – (mg/L nonsettling matter).

2.5 Turbidity
Nephelometric Method

A. General

1. References: See *Standard Methods* (1989, pp. 2–11 through 2–16) and EPA (1983, Method 180.1).
2. Outline of Method:
   The method compares the intensity of light scattered by the sample with the intensity of light scattered by a standard reference sample (formazine polymer suspension) under defined conditions. The higher the intensity of light scattered at 90° to the path of incident light, the higher the turbidity. Turbidity can be determined for any aqueous sample that is free of debris and rapidly settling materials. The most common interferences are dirty glassware, air bubbles in the sample, vibrations that disturb the surface of the sample, and color due to dissolved substances.

B. Special Reagents and Equipment

1. *Turbidimeter*: This is a nephelometer with a tungsten filament light source and one or more photoelectric detectors with a readout device (Hach Model 2000, 18900, or the equivalent).
2. *Sample Tubes*: These are clear colorless glass tubes to fit the turbidimeter.
3. *Turbidity-Free Water*: Use DDW from the laboratory purification system. If the turbidity of such water is greater than 0.05 nephelometric turbidity units (NTU), pass it through a 0.45 or 0.2 μm membrane filter to produce water of acceptable quality.
4. *Turbidity Standards*: Purchase the prepared formazine standards in sealed tubes from the manufacturer or consult the references listed under 2.5 A.1. for directions to make these.

C. Standardization

Note: If preparing standards, consult the references above.

1. Sealed standards from the manufacturer (0.61, 10, 100 and 1000 NTU) should be stored at room temperature. Handle carefully to avoid scratching glass surfaces.

2. Choose the standard or standards necessary to approximate the turbidity range of the samples.
3. Use laboratory tissues (i.e., Kimwipes) to clean the tubes thoroughly. The light path is directed through the bottom of the tubes. Never touch the bottom of the tubes.

   Note: DO NOT shake the standards. This causes excessive air bubbles and inaccurate standardization.

4. Place the standard in the turbidimeter chamber. Cover with the dark cylinder to keep stray light from interfering.
5. Turn the range switch to the appropriate setting to read the standard. Adjust, standardize, or calibrate knob until the standard reads correctly. Be sure to read on the appropriate meter scale (e.g., calibrate the 10 NTU standard on the 0–10 NTU range only).

   Note: Many turbidimeters require the use of a cell riser when using a 100 or 1000 NTU standard. However, these upper scales are used only to indicate the need for dilutions (see 2.5 D).

D. Procedure

1. Mix a total (unfiltered) sample thoroughly to disperse solids. When air bubbles disappear, pour the sample into a clean turbidity tube.
2. Dry the outside surface of the tube and place it in the turbidimeter.
3. If the reading is ≤ 40, read the turbidity in NTUs directly from the instrument scale.

   Note: Any time the range switch is changed, the instrument must be recalibrated with the appropriate standard. Record the range and standard used.

4. If the reading is >40, dilute the sample with turbidity-free water until the scale reading is less than 40 units. Record the dilution factor.

E. Calculations

1. If the reading of an undiluted sample is ≤ 40 units, report the results in NTUs as a direct reading of the instrument scale.
2. For a sample requiring dilution:

$$\text{Nephelometric Turbidity Units (NTU)} = \frac{A \times (B + C)}{C}$$

where

A = NTU found in diluted sample; instrument scale reading
B = Volume of turbidity-free dilution water in mL
C = Sample volume taken for dilution, mL

F. Notes

1. Report results as follows:

NTU	Record to Nearest
0.0 - 1.0	0.05
1 - 10	0.1
10 - 40	1
40 - 100	5
100 - 400	10
400 - 1000	50
> 1000	100

Methods for the Determination of Metals

3. Methods for the Determination of Metals

3.1 Calcium
EDTA Titrimetric Method

A. General

1. References: See *Standard Methods* (1989, pp. 3–85 through 3–87) and EPA (1983, Method 215.2).
2. Outline of Method:
   Hardness in water is primarily caused by free calcium and magnesium ions, both of which are readily complexed in the presence of ethylenediaminetetraacetic (EDTA) disodium salt. Magnesium precipitates as insoluble $Mg(OH)_2$ at a pH greater than 12.0 while the calcium ions present remain in solution; therefore, at pH = 12–13, only the calcium ions present in the sample combine with the EDTA titrant. Endpoint detection is facilitated by Calver II indicator, which is red in the presence of Ca^{++} and royal blue when all of the free calcium ions have been complexed by the EDTA. Alkalinity in excess of 300 mg/L as $CaCO_3$ may cause endpoint problems.

   Note: Because of the high pH used in this procedure, the titration should be performed immediately after the addition of the alkali, generally within 5 min.

B. Special Reagents

1. *Hydroxide Solution, 1 N*: Dissolve 56.1 g of KOH (or 40.00 g NaOH) in DDW and dilute to 1 L.
2. *0.01 M EDTA Titrant*: Dissolve exactly 3.723 g EDTA disodium salt ($Na_2H_2C_{10}O_8N_2 \cdot 2H_2O$) in DDW and dilute to 1 L. Store in a polyethylene or Pyrex bottle.
3. *Calver II Calcium Indicator*: This is manufactured by the Hach Chemical Co.
4. *Ammonia Hydroxide, 3 N*: Add 240 mL conc NH_4OH to about 700 mL of DDW and dilute to 1 L.
5. *Standard Calcium Solution*: Weigh 1.000 g primary standard anhydrous calcium carbonate, $CaCO_3$, into a 500 mL Erlenmeyer flask. Using a funnel, add a little at a time 1 + 1 HCl (5 mL HCl + 5 mL distilled water should make up enough) until all the $CaCO_3$ has dis-

solved. Add approximately 200 mL DDW and boil to expel the CO_2. Cool, (determine pH and adjust to ~5 by adding 3 N NH_4OH or 1 + 1 HCl as required). Transfer quantitatively to a 1 L volumetric flask. Rinse the Erlenmeyer three times with 50 mL quantities of DDW and add the rinse water to the volumetric flask. Dilute to 1 L with DDW.

$$1.00 \text{ mL} = 1 \text{ mg } CaCO_3$$

C. Standardization

1. Add 5.0 mL of the standard calcium solution to a 125 mL Erlenmeyer flask. Add 45 mL DDW by graduated cylinder.
2. Add 2 mL of the 1 N hydroxide solution and 0.1 to 0.2 g of the Calver II calcium indicator.
3. Titrate with the EDTA disodium salt titrant, until the color changes from red to blue. Standard solutions should take approximately 5 mL of titrant.
4. Set up two 50 mL DDW blanks and follow the same procedure for reagent addition and titration as described for the standards.

D. Procedure

1. Measure a 50 mL sample into a 125 mL Erlenmeyer flask. Make dilutions when necessary so that titrant volumes are greater than 1 mL and less than 15 mL.
2. Add 2 mL of the 1 N hydroxide solution, or a volume sufficient to produce a pH of 12–13 in the 50 mL sample. Add 0.1 to 0.2 g of Calver II calcium indicator.
3. Titrate slowly with the EDTA disodium salt solution (0.01 M) until the color changes from red to blue. Wait 30–60 sec and titrate to blue again. (Complete this second step only once). 1 mL of titrant (0.0100 M EDTA) is equivalent to exactly 400.8 μg Ca^{++}. Record sample size, dilution, and mL of the titrant used to reach the endpoint.

E. Calculation

Calcium as Ca^{++}:

$$mg\ Ca^{++}/L = \frac{(A - B) \times D \times 400.8}{mL\ sample}$$

Calcium hardness as $CaCO_3$:

$$mg\ CaCO_3/L = \frac{(A - B) \times D \times 1000}{mL\ sample}$$

where

A = mL of titrant used for the sample
B = mL of titrant used for the blank
C = mL of titrant used for standard
D = mg $CaCO_3$ equivalent to 1.00 mL EDTA titrant

$$= \frac{5\ mg\ CaCO_3\ std}{(C - B)} \approx 1$$

F. Notes

For increased sensitivity:

1. Use 0.004 M EDTA titrant.
2. Use Eriochrome Blue Black R indicator.
   a. Use 200 mg indicator/100 g NaCl.
   b. Use enough indicator to give a good red color (0.2 - 0.4g).
3. Larger samples may also be used (100 to 1000 mL).

3.2 Hardness
EDTA Titrimetric Method

A. General

1. References: *Standard Methods* (1989, pp. 2–52 through 2–57), and EPA (1983, Method 130.2).
2. Outline of Method:
   The hardness of water is due mainly to the presence of Ca^{++} and Mg^{++}. These ions form a chelated soluble complex in the presence of ethylenediaminetetraacetic acid (EDTA). Eriochrome Black T indicator is used, and as the EDTA is added the solution will turn from wine red to blue at the endpoint. The sharpness of the endpoint is pH dependent (pH 10.0 ± 0.1). To minimize chances of $CaCO_3$ precipitation, the titration should take no more than 5 min. The indicator functions best at room temperature.

B. Special Reagents

1. *Buffer Solution*: Dissolve 1179 g EDTA disodium salt and 0.780 g $MgSO_4 \cdot 7H_2O$ in 50 mL DDW. Add this solution to 16.9 g NH_4Cl and 143 mL conc NH_4OH with mixing and dilute to 250 mL with DDW. Store in tightly stoppered plastic or resistant-glass container. It will be stable less than 1 month.
2. *Inhibitors*: If needed, see *Standard Methods.*
3. *Eriochrome Black T Indicator*: Mix together 0.5 g dye and 4.5 g hydroxylamine hydrochloride. Dissolve this mixture in 100 mL of 95% ethyl or isopropyl alcohol.
4. *Standard EDTA Titrant, 0.01 M*: See 3.1.
5. *Standard Calcium Solution*: See 3.1.

C. Standardization

1. Set up standards and blanks as described for calcium (prior to reagent addition).
2. Add 1–2 mL buffer and 1–2 drops of the indicator solution. Titrate with the standard EDTA titrant. Standards should take approximately 5 mL of titrant and blanks $\leq$0.5 mL.

D. Procedure

1. Measure a 50 mL sample (or aliquots diluted to 50 mL) into a 125 mL Erlenmeyer flask. Add 1–2 mL buffer. Usually 1 mL of buffer will be sufficient to give a pH of 10.0–10.1. Make dilutions when necessary so that titrant volumes are greater than 1 mL and less than 15 mL.
2. Add 1–2 drops of indicator and titrate slowly, stirring continuously, until the last reddish tinge disappears from the solution (adding the last few drops at 3–5 sec intervals). 1 mL 0.0100 M EDTA should be equivalent to 1 mg $CaCO_3$.

E. Calculations

1. Total hardness as $CaCO_3$

$$\text{mg } CaCO_3/L = \frac{(A - B) \times D \times 1000}{\text{mL sample}}$$

where

A = mL of titrant used for the sample
B = mL of titrant used for the blank
C = mL of titrant used for standard
D = mg $CaCO_3$ equivalent to 1.00 mL EDTA titrant

$$= \frac{5 \text{ mg } CaCO_3}{(C - B)} \approx 1$$

2. Magnesium
   a. Calculate magnesium hardness by subtraction

   Magnesium Hardness (mg $CaCO_3$/L) =
   Total Hardness (mg $CaCO_3$/L) – Calcium Hardness (mg $CaCO_3$/L)

   b. Calculate magnesium as Mg^{+2}:

   mg Mg^{+2}/L = Magnesium Hardness as mg $CaCO_3$/L × 0.244

3.3 Hexavalent Chromium Colorimetric Method

A. General

1. Reference: See *Standard Methods* (1989, pp. 3–90 through 3–93).
2. Outline of Method:
   Dissolved hexavalent chromium, in the absence of interfering amounts of substances such as molybdenum, vanadium, and mercury, may be determined colorimetrically by reaction with diphenylcarbazide in an acidic solution. A red–violet color complex is produced that can be measured at 540 nm in determining the hexavalent chromium in solution.
3. See *Standard Methods* (1989, pp. 3–90 through 3–93) for methods to determine total chromium.

B. Special Reagents

1. *Stock Chromium Solution*: Dissolve 141.4 mg $K_2Cr_2O_7$ in DDW and dilute to 1 L. 1.00 mL = 50.0 μg Cr^{+6}.
2. *Nitric Acid*: Use conc HNO_3.
3. *Diphenylcarbazide Solution*: Dissolve 250 mg 1,5-diphenylcarbazide in 50 mL acetone. Store in a brown bottle. Prepare fresh as needed (e.g., as solution becomes discolored).

C. Standardization

1. Dilute 10.0 mL stock chromium solution up to 100 mL with DDW.

   $$1.00 \text{ mL} = 5\ \mu\text{g Cr}^{+6}\text{, or } 5000\ \mu\text{g/L solution}$$

2. Dilute 1.00 mL of the 5000 $\mu gCr^{+6}/L$ solution up to 100 mL, with DDW.

   $$1.00 \text{ mL} = 0.05\ \mu\text{gCr}^{+6}\text{/L or } 50\ \mu\text{g/L standard solution}$$

3. Using the 50 μg/L standard solution, set up the following dilutions for a standard curve.

Final Concentration μg/L	mL of 50 μg/L Standard Solution
0.5	1.0 mL diluted up to 100 mL
1.0	2.0 mL diluted up to 100 mL
2.0	4.0 mL diluted up to 100 mL
8.0	16.0 mL diluted up to 100 mL

Treat the standards the same as the samples in the following procedure.

D. Procedure

1. Pipet 25 mL of the sample or standard into a 125 mL Erlenmeyer flask.
2. Add 1–2 drops HNO_3 to each of the samples. pH should be 1.0 ± 0.3.
3. Add 0.50 mL diphenylcarbazide to each sample using an automatic pipet (do not use a serological pipet). Allow 5 to 10 min for full color development.
4. Read the absorbance against DDW at 540 nm, using a 5 cm cell.

E. Calculation

Calculate the hexavalent chromium in the sample as follows:

$$\mu g\ Cr^{+6}/L = (Abs_{sample} - Abs_{blank})\ (m^{-1})(df)$$

where

$$m^{-1} = \frac{\Delta\ concentration}{\Delta\ absorbance} = \frac{1 - 0}{Abs_{std} - Abs_{blank}}$$

$$\text{or} \quad \frac{2 - 0}{Abs_{std} - Abs_{blank}}$$

$$\text{or} \quad \frac{8 - 0}{Abs_{std} - Abs_{blank}}$$

df = Dilution factor

If a 10% sample is used: df = 10; if a 5% sample is used: df = 20.

The concentration of the hexavalent chromium in a sample may also be calculated using a linear regression of the calibration standards; or plot the absorbance of the calibration standards against the calibration concentrations and compute the sample concentration directly from the linear standard curve.

3.4 Preparation of Samples for the Determination of Metals Atomic Absorption (AA) or Inductively Coupled Plasma Emission Spectroscopy (ICP) Methods

A. General

1. References: See Parker (1972), *Standard Methods* (1989, pp. 3–5 through 3–11), and EPA (1983, Methods 200.0 and 200.7).
2. Outline of Method:
   Metals concentrations in water samples can be determined by atomic absorption spectrophotometry or inductively coupled plasma emission spectroscopy. The analytical processes used include absorption or emission techniques, the flameless or carbon furnace technique, and special procedures such as cold vapor (mercury determination) and hydride generation (arsenic and selenium). Some metals can be determined by either a flame, plasma, or flameless technique. In general, the carbon furnace technique results in the determination of lower concentrations by increased sensitivity and lower detection limits.

B. Special Reagents

1. *1 + 1 Nitric Acid, HNO_3*: Slowly and carefully add 500 mL conc HNO_3 to 500 mL DDW. Mix carefully and cool.

C. Types of Samples

1. *Dissolved Metals*: Metals that pass through a 0.45 μm membrane filter.
2. *Suspended Metals*: Metals retained by a 0.45 μm membrane filter.
3. *Total Metals*: Metals in both the dissolved and suspended portions of the sample.

Trace metals are commonly found in a dissolved state. In general, depending on the kind of sample, fewer types of metals are in an organic or particulate form. Table 3.1 lists the results of a comparison of suspended and dissolved metals.

Table 3.1 Comparison of Suspended and Dissolved Trace Metals in Surface Waters. (Samples are from U.S. rivers, collected over a five-year period)

	Suspended Metals[a]		Dissolved Metals[b]	
	Frequency of Detection, %	Mean, μg/L	Frequency of Detection, %	Mean, μg/L
Ag	0	. . .	7	2.6
Al	97	3,860	31	74
As	. . .	. . .	5.5	64
Ba	95	38	99.4	43
Be	18	0.34	5	0.19
Cd	0	. . .	3	9.5
Co	0	. . .	3	17
Cr	8	30	25	9.7
Cu	62	26	74	15
Fe	100	3,000	76	52
Mn	93	105	51	58
Mo	. . .	. . .	33	68
Ni	3	29	16	19
Pb	2	120	19	23
Sr	10	58	100	217
V	0	. . .	3	40
Zn	64	62	76.5	64

Source: Kopp and Kramer, 1967. Used with permission.
[a] 288 samples.
[b] 1,577 samples.

D. Procedure

1. Use clean, polyethylene containers for samples. When possible, soak containers up to 48 hr with 1 + 1 HNO_3 and rinse well with DDW; otherwise, rinse containers with 1 + 1 HNO_3 and then rinse well with DDW.
2. *Dissolved Metals*: Filter sample through a 0.45 μm membrane filter. Preserve filtrate with 1 + 1 HNO_3. Usually, 3 mL acid per liter of sample is sufficient to lower the pH to ≤2.

   Note: For dissolved metals, always filter sample BEFORE preserving with acid.

   For highly alkaline samples, add more acid if necessary to lower pH to ≤2.
   Suspended Metals: Filter sample with a 0.45 μm membrane filter. Retain the filter and follow the procedure given in Parker (1972).
   Total Metals: Preserve unfiltered sample with 1 + 1 HNO_3, adding enough to obtain a final pH of ≤2.
3. The total metals samples must undergo an acid digestion before analysis. The digestion process is run by the analyst responsible for the

atomic absorption instrument. The HNO_3 digestion technique is one of several available. See Parker (1972) and *Standard Methods* (1989).

4. Sample volumes required:
   Dissolved Metals: 10 mL for each metal determined by ordinary flame, flameless or plasma technique.
   100 mL for mercury
   50 mL for arsenic
   50 mL for selenium
   Total Metals: 100 mL in addition to the volumes listed above.

E. Special Samples

For samples other than water, or for the determination of very low concentrations of metals in water, there are procedures available for extractions, digestions, and concentrating techniques. See Parker (1972), *Standard Methods* (1989), and EPA (1983).

Methods for the Determination of Inorganics and Nonmetals

4. Methods for the Determination of Inorganics and Nonmetals

4.1 Acidity
Titrimetric and Potentiometric Methods

A. General

1. References: See *Standard Methods* (1989, pp. 2–30 through 2–34) and EPA (1983, Methods 305.1 and 305.2).
2. Outline of Method:
   The acidity of a water is a measure of the concentration of acids (both weak and strong) that react with a strong base. Acidity may be expressed as mg $CaCO_3$/L or as meq/L. Sources of acidity are strong mineral acids, weak acids and ferrous (Fe^{+2}) or other polyvalent cations in a reduced state. The sample pH is lowered by addition of standard acid (if necessary), treated for metal ions with peroxide, and titrated with standard base to the phenolphthalein endpoint (pH 8.3).

B. Special Reagents

1. *Carbon Dioxide-Free Water*: Boil fresh DDW from the purification system for 15 min. Cover and cool to room temperature (pH ≥ 6.0 and the conductivity $< 2\mu$mhos/cm). Use this water to prepare all reagents and standards for making dilutions.
2. *Potassium Hydrogen Phthalate (KHP) Solution, Approximately 0.05 N*: Dry 15–20 g of primary standard grade KHP at 120°C for 2 hr. Cool in a desiccator. Weigh out 10.0 ± 0.5 g to the nearest mg. Transfer quantitatively to a volumetric flask, dissolve in CO_2-free DDW, and dilute to 1 L. Transfer to a tightly stoppered glass reagent bottle and indicate the quantity of KHP per liter on the label.
3. *Stock Sodium Hydroxide, Approximately 0.1 N*: Dissolve 4 g of sodium hydroxide (NaOH) in about 50 mL of CO_2-free DDW with stirring. Cool and transfer quantitatively to a volumetric flask and dilute to 1 L with DDW. 1 mL $\cong$ 5 mg $CaCO_3$. Store in a polyethylene bottle with a tight polyethylene screw cap.
4. *Standard Sodium Hydroxide Titrant, Approximately 0.02 N*: Place 200 mL of 0.1 N NaOH in a volumetric flask and dilute to 1 L with CO_2-free DDW. Standardize against KHP (see 4.1 C) and record exact normality

on the reagent bottle. Store in a polyethylene bottle with a tight screw cap. It may be necessary to protect from atmospheric CO_2 by installing a drying tube and syphon (see 1.11).

5. *Stock Sulfuric Acid (H_2SO_4), Approximately 0.1 N*: For preparation see 4.2, B.2.
6. *Standard Sulfuric Acid, Approximately 0.02 N*: Place 200 mL of 0.1 N H_2SO_4 in a volumetric flask and dilute to 1 L with DDW. Record exact normality on the reagent bottle.
7. *Hydrogen Peroxide (H_2O_2), 30%*: Store in the refrigerator. Caution: H_2O_2 is a fairly strong oxidizing agent. Handle with care.
8. *Sodium Thiosulfate, 0.1 N*: Dissolve 25 g $Na_2S_2O_3 \cdot 5H_2O$ in DDW and dilute to 1 L.

C. Standardization

Note: Use CO_2-free DDW at all times.

1. Standardization of 0.02 N NaOH titrant:
   a. Pour two blanks using 50 mL of DDW and pipet duplicate 10.0 mL aliquots of the 0.05 N KHP solution plus 40 mL DDW by graduated cylinder into 125 mL beakers. Insert calibrated pH electrodes and titrate to pH 8.7. The KHP solutions will require about 25 mL of titrant.
   b. Calculate the actual normality of the NaOH titrant:

$$\text{Normality} = \frac{A \times B}{204.2 \times (C - D)} \approx 0.02$$

where

A = g KHP/L weighed out to make ~0.05 N KHP
B = mL ~0.05 N KHP used
C = mL NaOH titrant used for KHP solution
D = mL NaOH titrant used for blank

Note 1: The 0.02 N NaOH titrant should be standardized each time the analysis is run as it will degrade.
Note 2: The average blank value (D) will be used again in the sample calculations.

2. Standardization of 0.02 N H_2SO_4: Refer to 4.2, B.2 to standardize the 0.1 N H_2SO_4 titrant. Prepare 0.02 N H_2SO_4 from the solution. Multiply the normality calculated above by 0.2 to get the actual normality of the 0.02 N H_2SO_4.

D. Procedure

1. Measure 50 mL of total (unfiltered) sample (or an aliquot diluted to 50 mL) into a 125 mL beaker. Insert a calibrated pH probe and magnetic stir bar. Record the pH of the sample. If residual free chlorine is present, add one drop of 0.1 N sodium thiosulfate.
2. If the sample pH is above 4.0, add 5.0 mL increments of 0.02 N H_2SO_4 to lower the pH to 4.0 or less. Record volume of acid added.
3. Remove the electrodes. Add 5 drops of 30% H_2O_2 if the sample is known or suspected to contain hydrolyzable metal ions or reduced forms of polyvalent cations (i.e. Fe^{+2}). Such samples would include industrial wastes and those from acid mine drainages. Boil the sample for 2–5 min. Cool to room temperature.
4. Titrate with 0.02 N NaOH to pH 8.3 adding titrant about 1 mL at a time and slowing down near the endpoint. Record the volume of base added.

E. Calculations

$$\text{Acidity, as mg CaCO}_3\text{/L} = \frac{[(A - B) \times C] - [(D \times E)] \times 50{,}000}{\text{mL sample}}$$

where

- A = mL NaOH titrant used for sample
- B = mL standard NaOH titrant used for blank
- C = actual normality of standard NaOH titrant
- D = mL standard H_2SO_4 used (*Note*: This term may be zero if pH of sample is ≤ 4.0.)
- E = actual normality of standard H_2SO_4

If the analyst is to report the acidity in millequivalents per liter, divide the acidity (mg $CaCO_3$/L) values by 50.

$$\text{Acidity as meq/L} = \frac{\text{Acidity as mg CaCO}_3\text{/L}}{50}$$

4.2 Alkalinity
Potentiometric and Colorimetric Methods

A. General

1. References: See *Standard Methods* (1989, pp. 2-35 through 2-39) and EPA (1983, Method 310).
2. Outline of Method:
   Alkalinity measures the ability of water to accept protons. The total alkalinity generally is imparted by the bicarbonate, carbonate, and hydroxide components dissolved in the water and may be measured by titration with acid. Samples should be collected in polyethylene and analyzed within 24 hr.

B. Special Reagents

1. *Sodium Carbonate Solution, 0.05 N*: Dry 2 to 3 g primary standard Na_2CO_3 at 250°C for 4 hr and cool in a desiccator. Weigh 1.25 g ± 0.1 g (to the nearest mg), transfer into a 500 mL volumetric flask and fill the flask to the mark with DDW. The solution is generally good for only 1 week.
2. *Standard Sulfuric Acid or Hydrochloric Acid, 0.1 N*: Dilute 3.0 mL conc H_2SO_4 or 8.3 mL conc HCl to 1 L with DDW. Standardize against 40.00 mL 0.05 N Na_2CO_3 solution with about 60 mL DDW by titrating potentiometrically to a pH of about 5. Lift out electrodes, rinse into the beaker, and boil gently for 3 to 5 min under a watch glass cover. Cool to room temperature, rinse cover glass into beaker, and finish titrating to pH 4.5. Calculate normality:

$$\text{Normality, N} = \frac{M \times S}{26.50 \times C}$$

   where

   M = g Na_2CO_3 weighed into the 500 mL flask
   S = mL Na_2CO_3 solution taken for titration
   C = mL acid used

3. *0.02 N Sulfuric Acid or Hydrochloric Acid*: Dilute 200.00 mL 0.1000 N Standard Acid to 1 L with DDW and standardize (4.2, C.).

Table 4.1 Alkalinity Color Indicators and pH Endpoint Data.

Indicator	pH at Equivalence Point	Color Change	mL Indicator/100 mL Sample
Phenolphthalein	8.3	pink → clear	0.1
Mixed Indicator	5.0	greenish blue → light blue with lavender gray	0.15
	4.8	→ light pink-gray with bluish cast	
	4.6	→ light pink	
Methyl Orange	4.6	yellow → orange	0.5
	4.0	→ pink	

4. *Phenolphthalein Indicator*: Dissolve 0.5 g phenolphthalein disodium salt in DDW and dilute to 100 mL.
5. *Mixed Indicator*: Dissolve 20 mg methyl red and 100 mg bromcresol green sodium salt into 100 mL of 95% ethanol or DDW.
6. *Methyl Orange Indicator*: Dissolve 50 mg of methyl orange powder in DDW and dilute to 100 mL.

C. Standardization

1. Pipet 10.0 mL of 0.05 N Na_2CO_3 into a 125 mL beaker. Add 90 mL DDW. (Do two replicates.)
2. Titrate with 0.02 N H_2SO_4 to a 4.5 pH endpoint (colorimetric or pH meter). Approximately 24 mL of titrate will be used. In the absence of interfering color or turbidity, color indicators may be used to detect the equivalence point. Whenever an indicator is used, titrate against a white background. (See Table 4.1.)
3. Repeat the above procedure using 100 mL of DDW as a blank. Calculate the normality of the acid.

$$\text{Normality} = \frac{M \times 10.0}{26.50 \times (C - B)}$$

where

M = g Na_2CO_3 weighed into the 500 mL flask
C = mL acid used for the standard
B = mL acid used for the blank

Table 4.2 Volume of Sample, Titrant Normality, and Approximate pH Endpoint for Various Alkalinity Concentrations.

GENERAL ENDPOINT CHART		APPROXIMATE FACTOR CHART	
Total Alk. Range mg/L	pH	Sample Volume mg/L	Acid N
0–25	5.2	50	0.001
25–50	5.1	100	0.002
50–90	5.0	50	0.002
90–130	4.9	100	0.02
130–210	4.8	50	0.02
210–330	4.7	50	0.02
330–445	4.6	50	0.02
445–	4.5	25	0.02

D. Procedure

1. To predetermine total alkalinity endpoint pH:
   a. Measure out 100 mL of the sample into a 250 mL beaker and titrant using 0.02 N H_2SO_4 adding titrant slowly, 1 mL at a time, to a pH of about 4.5. Continuously mix on a magnetic stirrer while titrating.
   b. Plot results pH vs mL titrant. Resultant endpoint pH (about 4.5) will then be used for subsequent samples of that water type.
   c. An example of typical results is shown in Table 4.2. This can be used when time does not permit a titration curve.
   d. Dilute titrant as necessary for lower alkalinities. Use a separate buret for diluted titrant.
2. Alkalinity:
   a. Measure 100 mL of unfiltered sample into a beaker. Use sample volumes less than 100 mL if the alkalinity is high and would require more than 25 mL of titrant. Insert bar magnet and pH meter electrodes; switch on magnetic stirrer making sure the bar magnet does not touch the electrodes. Alternately, use color indicators.
   b. Record initial pH, sample volume, and sample temperature.
   c. If initial pH is above 8.3, titrate to 8.3 using pH meter or indicator, recording mL 0.02 N H_2SO_4 used. This is phenolphthalein endpoint.
   d. Titrate to a pH of 4.5 or appropriate endpoint using pH meter or indicator and recording volume of titrant. This is the total alkalinity endpoint.

Table 4.3 Alkalinity Relationships.

Result of Titration	Hydroxide Alkalinity as $CaCO_3$	Carbonate Alkalinity as $CaCO_3$	Biocarbonate Alkalinity as $CaCO_3$
P = 0	0	0	T
P < 1/2 T	0	2P	T - 2P
P = 1/2 T	0	2P	0
P < 1/2 T	2P - T	2(T - P)	0
P = T	T	0	0

Source: Standard Methods for the Examination of Water and Wastewater (17th edition). Copyright 1989 by APHA, AWWA, and WPCF. Reprinted with permission.
Note: P = Phenolphthalein Alkalinity, T = Total Alkalinity.

E. Calculation

$$\text{Alkalinity, mg/L as CaCO}_3 = \frac{(A - B) \times N \times 50{,}000}{\text{mL sample}}$$

where

A = mL standard acid used for sample
B = mL standard acid used for blank
N = normality of acid

F. Notes on Alkalinity Composition (see Table 4.3)

Conversion of total alkalinity to inorganic carbon:
See Table 4.4 and select the conversion factor from the sample pH and temperature taken at the time of the alkalinity analysis.

$$\text{mg C/L} = (\text{Total Alkalinity})\ (\text{Conversation Factor})$$

Table 4.4 Factors for the Conversion of Total Alkalinity to Milligrams of Carbon Per Liter.

			Temperature (°C)			
pH	0	5	10	15	20	25
5.0	9.36	8.19	7.16	6.55	6.00	5.61
5.1	7.49	6.55	5.74	5.25	4.81	4.51
5.2	6.00	5.25	4.61	4.22	3.87	3.63
5.3	4.78	4.22	3.71	3.40	3.12	2.93
5.4	3.87	3.40	3.00	2.75	2.53	2.38
5.5	3.12	2.75	2.43	2.24	2.06	1.94
5.6	2.53	2.24	1.98	1.83	1.69	1.59
5.7	2.06	1.83	1.62	1.50	1.39	1.31
5.8	1.69	1.50	1.34	1.24	1.15	1.09
5.9	1.39	1.24	1.11	1.03	0.96	0.92
6.0	1.15	1.03	0.93	0.87	0.82	0.78
6.1	0.96	0.87	0.77	0.73	0.70	0.67
6.2	0.82	0.74	0.68	0.64	0.60	0.58
6.3	0.69	0.64	0.59	0.56	0.53	0.51
6.4	0.60	0.56	0.52	0.49	0.47	0.45
6.5	0.53	0.49	0.46	0.44	0.42	0.41
6.6	0.47	0.44	0.91	0.40	0.38	0.37
6.7	0.42	0.40	0.38	0.37	0.35	0.35
6.8	0.38	0.37	0.35	0.34	0.33	0.32
6.9	0.35	0.34	0.33	0.32	0.31	0.31
7.0	0.33	0.32	0.31	0.30	0.30	0.29
7.1	0.31	0.30	0.29	0.29	0.29	0.28
7.2	0.30	0.29	0.28	0.28	0.28	0.27
7.3	0.29	0.28	0.27	0.27	0.27	0.27
7.4	0.28	0.27	0.27	0.26	0.26	0.26
7.5	0.27	0.26	0.26	0.26	0.26	0.26
7.6	0.27	0.26	0.26	0.25	0.25	0.25
7.7	0.26	0.26	0.25	0.25	0.25	0.25
7.8	0.25	0.25	0.25	0.25	0.25	0.25
7.9	0.25	0.25	0.25	0.25	0.25	0.25
8.0	0.25	0.25	0.25	0.25	0.24	0.24
8.1	0.25	0.25	0.24	0.24	0.24	0.24
8.2	0.24	0.24	0.29	0.24	0.24	0.24
8.3	0.24	0.24	0.24	0.24	0.24	0.24
8.4	0.24	0.24	0.24	0.24	0.24	0.24
8.5	0.24	0.24	0.24	0.24	0.24	0.24
8.6	0.24	0.24	0.24	0.24	0.24	0.24
8.7	0.24	0.24	0.24	0.24	0.24	0.24
8.8	0.24	0.24	0.24	0.24	0.23	0.23
8.9	0.24	0.24	0.23	0.23	0.23	0.23
9.0	0.24	0.23	0.23	0.23	0.23	0.23
9.1	0.23	0.23	0.23	0.23	0.23	0.23
9.2	0.23	0.23	0.23	0.23	0.23	0.23
9.3	0.23	0.23	0.23	0.22	0.22	0.22
9.4	0.23	0.23	0.22	0.22	0.22	0.22

Source: Saunders et al. (1962). Used with permission.

4.3 Calcium Carbonate Saturation Langelier Saturation Index

A. General

1. References: See *Standard Methods* (1989, pp. 2–40 through 2–52), Deberry et al.(1982), and Singley et al.(1984).
2. Outline of Method:
   For a water containing calcium and bicarbonate, the pH at which the solution is saturated with calcium carbonate ($CaCO_3$) is the pH of saturation or pHs. The Langelier Saturation Index (SI) is a calculated value (actual pH-pHs) that gives an indication of whether $CaCO_3$ may be deposited (index >0) or dissolved (index <0). The index is not directly related to corrosion but indicates potential corrosion of piping materials. To determine the SI, the following data are required: total alkalinity mg/L as $CaCO_3$, calcium concentration, mg/L as $CaCO_3$, pH, temperature and total dissolved solids, mg/L.

B. Special Reagents

1. *Total Alkalinity*: See 4.2.
2. *Calcium*: See 3.1.
3. *pH*: Use a pH meter and calibration standards.
4. *Temperature*: Use a calibrated thermometer.
5. *Total Dissolved Solids*: See 2.3.

C. Standardization

Standardization is only required on the above determinations. The saturation index is based upon a calculation as defined by:

$$SI = pH - (A + B - \text{Log}\,[Ca^{++}] - \text{Log}\,[Alk])$$

where

pH = hydrogen ion concentration, pH units

A = $pk_2 - pk_s$ and is a function of water temperature (k_2 = second dissociation constant of H_2CO_3, and k_s = solubility constant of $CaCO_3$)

B is a function of total dissolved solids (ionic strength correction and conversion factors).

Ca^{++} = calcium hardness, mg/L as $CaCO_3$

Alk = total alkalinity, mg/L as $CaCO_3$

D. Procedure

1. Determine the total alkalinity as mg/L $CaCO_3$, calcium as mg/L $CaCO_3$, pH, temperature and total dissolved solids, mg/L on the water sample.

E. Calculations

Using Tables 4.5 and 4.6, values for A and B, calculate the SI by the following equation:

Table 4.5 Constant A as a Function of Water Temperature.

Sample Temperature (Degrees° C)	A
0	2.60
4	2.50
8	2.40
12	2.30
16	2.20
20	2.10
25	2.00
30	1.90
40	1.70
50	1.55
60	1.40
70	1.25

Table 4.6 Constant B as a Function of Total Dissolved Solids.

Total Dissolved Solids mg/L	B
0	9.70
100	9.77
200	9.83
400	9.86
800	9.89
1000	9.90

$$SI = pH - (A + B - \text{Log}\,[Ca^{++}] - \text{Log}\,[Alk])$$

Calculation Example:

For a water sample with the following characteristics:

Alkalinity	100 mg/L as $CaCO_3$ characteristics:
Calcium	80 mg/L as $CaCO_3$
Total Dissolved Solids	150 mg/L
Temperature	18°C
pH	7.6

calculate the Saturation Index (SI)

$SI = pH - (A + B - \text{Log}\,[Ca^{++}] - \text{Log}\,[Alk])$
for a temperature of 18°C

A = 2.15 (interpolation from Table 4.5, value midway between temperatures of 16 and 20°C), and

B = 9.80 (interpolation from Table 4.6, value midway between the values of 100 and 200 mg/L)

thus

$SI = 7.6 - (2.15 + 9.80 - \text{Log } 80 - \text{Log } 100)$

$SI = 7.6 - (2.15 + 9.80 - 1.90 - 2.0)$

$SI = 7.6 - 8.05$

$SI = - 0.45$

The water sample is undersaturated with respect to $CaCO_3$.

4.4 Aggressive Index

A. General

1. References: See Deberry et al. (1982) and Singley et al. (1984).
2. Outline of Method:
   The Aggressive Index (AI) was developed to assist consulting engineers in the selection of the proper pipe for water conveyance. The AI is based upon pH and the solubility of $CaCO_3$. It is essentially a simplified form of the Langelier Saturation Index (SI). To determine the AI, pH, alkalinity, and calcium hardness are required.

B. Special Reagents

1. *Total Alkalinity*: See 4.2.
2. *Calcium*: See 3.1.
3. *pH*: Use a pH meter and calibration standards.

C. Standardization

Standardization is only required on the above determinations. The AI is defined as follows:

$$AI = pH + \text{Log}\,[(Alk)(H)]$$

where

pH = hydrogen ion concentration, pH units
Alk = total alkalinity, mg/L as $CaCO_3$
H = calcium hardness, mg/L as $CaCO_3$

The values obtained are defined as follows:

$AI < 10$, highly aggressive water (corrosive)
$AI = 10–12$, moderately aggressive water
$AI > 12$, nonagressive water

D. Procedure

1. Determine the total alkalinity as mg/L $CaCO_3$, calcium as mg/L $CaCO_3$ (calcium hardness) and pH on the water sample.

E. Calculations

$$AI = pH + Log\,[(Alk)(H)]$$

Calculation example for a water sample with the following characteristics:

Alkalinity	100 mg/L as $CaCO_3$
Calcium hardness	80 mg/L as $CaCO_3$
pH	7.6

Calculate the Aggressive Index (AI).

$$AI = pH + Log\,[(Alk)(H)]$$
$$AI = 7.6 + Log\,[(100)(80)]$$
$$AI = 7.6 + Log\,[8000]$$
$$AI = 7.6 + 3.9$$
$$AI = 11.5$$

In this example, the water would be classified as moderately aggressive.

4.5 Boron
Carmine Method

A. General

1. Reference: See *Standard Methods* (1989, pp. 4-7 through 4-11).
2. Outline of Method:
   A solution of carmine in concentrated sulfuric acid will react with boron in aqueous solution. The original reagent color will change from a bright red to a bluish red or blue depending upon the concentration of boron in the sample. The carmine method is used for boron concentration in the range 100 μg/L–500 μg/L. Store samples in plastic. *It should be noted that the analyst is using concentrated acids and needs to adhere to rigorous safety precautions.*

B. Special Reagents

Store all reagents in polyethylene or other plastic boron-free containers.

1. *Hydrochloric Acid*: Use conc HCl.
2. *Sulfuric Acid*: Use conc H_2SO_4.
3. *Carmine Reagent*: Dissolve 460 mg carmine (Alum Lake) 40, in 1 L conc H_2SO_4. Allow to sit overnight for complete dissolution. Afterwards, store in the refrigerator.
4. *Stock Boric Acid for Standards*: Dissolve 0.5716 g anhydrous H_3BO_3 in DDW. Dilute up to 1 L.

C. Standardization

Use plastic volumetrics for the following standards:

1. From the stock boric acid solution:

Final Concentration μg B/L	Amount Stock Boric Acid Solution
1000	10.0 mL diluted up to 1L
1500	3.0 mL diluted up to 200 mL
3000	3.0 mL diluted up to 100 mL

2. From 1000 μg/L standard.

Final Concentration μg B/L	Amount 1000 μg/L Standard
100	10.0 mL diluted up to 1L
500	50.0 mL diluted up to 100 mL

3. Set up three DDW blanks and run duplicates on all standards. Treat the blanks and standards exactly as outlined below for the samples.

D. Procedures

1. Rinse plastic 150 mL beakers the day before with DDW and drain. Allow the carmine reagent to come to room temperature.
2. Set up duplicates for all samples. Pipet 2.0 mL of each sample or standard into a plastic beaker. Place each beaker on top of an ice bath. In a hood, carefully add to each sample 0.1 mL conc HCl and mix by swirling. Carefully add 10.0 mL conc H_2SO_4 by repipet to each sample mixing thoroughly after addition of the acid. Next add 10.0 mL of the carmine reagent to each sample by repipet and again mix thoroughly. IMPORTANT: Remove the repipet from the carmine reagent after adding it to the samples so that the carmine does not leach the boron out of the glass.
3. Allow 2 hr for a reaction period. The 1500 μg/L and 3000 μg/L standards will appear slightly blue. The color is stable for 2 hr after the reaction period.
4. Measure the absorbance at 585 nm in a 1 cm cell, using one of the blanks as the reference. Mix by swirling all samples and standards BEFORE pouring into the cuvette. Rinse cuvette three times with the sample. DO NOT RINSE IN BETWEEN SAMPLES WITH DDW. Rinse the acid from the outside of the cuvette with DDW and wipe dry with a Kimwipe.

   Pour one of the three sample blanks in the back cell of the spectrophotometer and allow the blank reading to stabilize initially and set the instrument to zero. Place the lowest blank in the back cell and record the readings of the other two blanks. Then read the standards and samples. To avoid error, make sure that no bubbles are present in the optical cell while making the photometric readings. The bubbles may appear as a result of incomplete mixing of the reagents. Check the calibration curve daily because the carmine reagent deteriorates.

Because of this, the absorbance of the standards varies somewhat. However, the 1000 μg/L standard can be expected to fall in the range of 0.036–0.058 at 610 nm.

E. Calculation

Calculate the concentration of B in a sample using a linear regression of the calibrations standards. Or plot the absorbance of the calibration standards against the standard concentrations and compute the sample concentration directly from the linear standard curve.

4.6 Chloride
Mercuric Nitrate Titrimetric Method

A. General

1. References: See *Standard Methods* (1989, pp. 4–67 through 4–70) and EPA (1983, Method 325.3).
2. Outline of Method:
   Chloride can be determined by the titration of an acidic sample with mercuric nitrate in the presence of an indicator. In the pH range of 2.3 to 2.8, diphenylcarbazone forms a purple complex with excess mercuric ions at the endpoint of the titration. The indicator-acidifier automatically adjusts the pH of most potable waters to pH 2.5 ± 0.1. Xylene cyanol FF facilitates endpoint detection, which goes from blue-green to blue to purple.
3. Mercury is classified as a hazardous waste. All mercury-containing waste from this test should be stored in a labeled waste container.

B. Special Reagents

1. *Indicator-Acidifier Reagent*: Dissolve in the following order 0.25 g s-diphenylcarbazone, 4.0 mL conc HNO_3, and 0.03 g xylene cyanol FF in 100 mL of 95% EtOH or isopropyl alcohol. Store in a dark bottle in the refrigerator.
2. *Mercuric Nitrate Titrant*: Dissolve 2.5 g $Hg(NO_3)_2 \cdot H_2O$ in 100 mL DDW containing 0.25 mL conc HNO_3.

   Note: Do not use mercuric nitrate crystals that have taken on water and appear wet. Keep this chemical stored in a desiccator and in the dark.

   Dilute to 1 L with DDW. Store in a dark bottle.

   Normality = 0.0141:
   1.00 mL is equivalent to 0.50 mg Cl^-

3. *Sodium Chloride Standard*: Dissolve 0.8241 g NaCl (dried at 140°C) in DDW and dilute to 1 L.

   Normality = 0.0141:
   1.00 mL = 0.50 mg Cl^-

C. Standardization

1. Measure 5.0 mL of the concentrated sodium chloride and 10 mg $NaHCO_3$ into each of two 250 mL Erlenmeyer flasks. Add 95 mL DDW by graduated cylinder. Pour two 100 mL blanks each containing 10 mg $NaHCO_3$.
2. Add 1.0 mL of the indicator-acidifier reagent, mix, and titrate with the mercuric nitrate titrant to a purple endpoint. Save this standard; the same specific hue of purple must be adopted for blanks and the samples.
3. Standards will take approximately 5 to 6 mL of titrant; blanks should take ≤ 1.0 mL titrant.

D. Procedure

1. Measure a 100 mL sample (or an aliquot diluted to 100 mL so that the chloride content is less than 10 mg) into a 250 mL Erlenmeyer flask.
2. Add 1 mL of the indicator-acidifier reagent; the pH at this point should be in the 2.3–2.8 range. At pH 2.5 ± 0.1 the indicator should give the sample a blue-green color.
3. Titrate with mercuric nitrate titrant to the same color as the standard.

E. Calculations

$$\frac{\text{mg Cl}^-}{\text{L}} = \frac{(A - B) \times N \times 35{,}450}{\text{mL sample}} = \frac{(A - B) \times \left[\frac{2.5}{(C - B)}\right]}{\text{mL sample}}$$

where

A = mL of titrant used for the sample
B = mL of titrant used for the blank
C = mL of titrant used for the standard

$$\text{Normality, N} = \frac{5 \times 0.5}{35{,}450 \times (C - B)} \approx 0.0141$$

F. Notes

1. HNO_3 in the indicator-acidifier is sufficient to neutralize a total alkalinity of 150 mg/L as $CaCO_3$ to the proper pH in a 100 mL sample. If the sample is alkaline, either dilute or neutralize (see *Standard Methods*, pp. 4–69).
2. $NaHCO_3$ is used in blanks and standards to assure proper pH upon addition of indicator-acidifier.

4.7 Residual Chlorine
Amperometric Titration Method

A. General

1. References: See *Standard Methods* (1989, pp. 4–45 through 4–58) and EPA (1983, Methods 330.1 and 330.2).
2. Outline of Method:
   Free available chlorine is titrated with standardized phenylarsine oxide titrant using amperometric endpoint detection at a pH between 6.5 and 7.5. At this pH range the combined chlorine reacts slowly. The combined chlorine can be titrated in the presence of the proper amount of potassium iodide in the pH range of 3.5–4.5.

B. Special Reagents

1. *Sodium Hydroxide, 0.3 N*: Dissolve 1.8 g of NaOH in 150 mL DDW.
2. *Standard Phenylarsine Oxide Titrant*: Dissolve ~0.8 g phenylarsine oxide powder, C_6H_5AsO, in 150 mL 0.3 N sodium hydroxide. After it settles, pour 110 mL of this solution into 800 mL DDW and mix thoroughly. Bring the solution to pH 6–7 with 6 N HCl solution and dilute to almost 1 L.

   Note: Make sure the volume is less than 1 L. Final adjustment will be made later. See 4.7 C.

   Caution: This solution is toxic; take care to avoid ingestion.

3. *Standard Sodium Arsenite, Approximately 0.1 N*: Accurately weigh a stoppered weighing bottle containing ~4.95 g primary standard grade arsenic trioxide, As_2O_3. Transfer quantitatively to a 1 L volumetric flask and again weigh the bottle. Do not attempt to brush out the remaining oxide. Moisten the As_2O_3 with DDW and add 15 g NaOH and 100 mL DDW. Swirl contents of the flask gently until the As_2O_3 is in solution. Dilute to about 250 mL with DDW and saturate the solution with CO_2 by bubbling with CO_2 gas. This will convert all the hydroxide to bicarbonate. Dilute to the 1 L mark, stopper the flask, and mix thoroughly.

   Caution: This solution is toxic; take care to avoid ingestion.

4. *Standard Iodine, 0.1 N*: Dissolve 40 g KI in 25 mL DDW, add 13 g resublimed iodine (I_2), and stir until dissolved. Transfer quantitatively to a volumetric flask and dilute to 1 L.
5. *Diluted Standard Iodine, 0.0282 N*: Dissolve 25 g KI in DDW in a 1 L volumetric flask, add the proper amount of standardized 0.1 N iodine solution to yield a 0.0282 N solution, and dilute to 1 L. Store in a brown bottle in the dark and standardize daily. (Keep the solution from all contact with rubber.)
6. *Phosphate Buffer Solution, pH 7*: Dissolve 25.4 g anhydrous potassium dihydrogen phosphate, KH_2PO_4, and 34.1 g anhydrous disodium hydrogen phosphate Na_2HPO_4, in 800 mL DDW. Add 2 mL sodium hypochlorite (NaOCl) solution containing 1% available chlorine and mix. Protect from sunlight for several days and then expose to sunlight until no residual chlorine remains. If necessary, use sodium sulfite for final dechlorination. Finally, dilute to 1 L with DDW and filter if necessary.
7. *Potassium Iodide Solution*: Dissolve 50 g potassium iodide, KI, and dilute to 1 L using freshly boiled and cooled DDW. Store in a brown glass-stoppered bottle in the refrigerator.
8. *Acetate Buffer Solution, pH 4*: Mix 480 g concentrated (glacial) acetic acid, $HC_2H_3O_2$, and 243 g sodium acetate trihydrate, $NaC_2H_3O_2 \cdot 3H_2O$ in 400 mL DDW and dilute to 1 L.

   Caution: Work with concentrated glacial acetic acid in a hood.

9. *Starch Indicator Solution*: See 4.20.

C. Standardization

$$\text{Normality, N, of sodium arsenite} = \frac{\text{g } As_2O_3}{49.455}$$

Standard Iodine, 0.1 N:

Measure 40.0 mL of 0.1 N sodium arsenite solution into a flask and titrate with the 0.1 N iodine solution, using starch solution as indicator. Accurate results are only obtained if the solution is saturated with CO_2. This can be done by bubbling CO_2 through the solution or by adding acid (dilute HCl) near the endpoint.

$$\text{Normality, N, Standard Iodine Solution} = \frac{40.0 \times 0.1}{\text{mL iodine solution}}$$

Standard Phenylarsine Oxide Titrant:

Determine the normality of phenylarsine oxide solution by amperometric titration of standard 0.0282 N iodine solution. Adjust the final concentration of phenylarsine oxide to the desired 0.00564 N before completing the final standardization with the amperometric titrator (i.e., 5 mL PAO = 1 mL I_2).

1.00 mL of exactly 0.00565 N phenylarsine oxide titrant = 200 mg available chlorine. If a 200 mL sample is used (marked on titrator), 1 mL PAO = 1 mg/L chlorine.

Note: Preserve phenylarsine oxide titrant with 1 mL chloroform. (When using chloroform, always work under a hood. Use safety glasses or a mask and wear gloves.)

D. Procedure

Sample Volume:	Residual chlorine	Sample Size, mL
	2 mg/L or less	200
	>2 mg/L	100 or proportionately less

Measure a sample volume such that no more than 2 mL of phenylarsine oxide titrant is necessary.

Free Available Chlorine:

1. If the pH of the sample is not between 6.5 and 7.5, add 1 mL pH 7 phosphate buffer solution to produce a pH of 6.5–7.5.
2. First adjust the meter until the needle is about in the middle of the range. Titrate with standard phenylarsine oxide (PAO), observing the needle deflection changes on the microammeter. For each increment of PAO titrant added, the needle moves a certain increment to the left. When the needle moves a smaller increment for the same increment of PAO titrant, the endpoint has been reached. This can be verified by adding another increment of PAO.

Combined Available Chlorine:

1. To the sample remaining from the free-chlorine titration, first add 1 mL KI solution, and then 1 mL acetate buffer solution.
2. Titrate with phenylarsine oxide titrant to an end point using the above procedure.

E. Calculation

Convert the individual titrations for free available and combined available chlorine into mg/L by the following equation:

$$\text{mg Cl as } Cl_2/L = \frac{A \times 200}{\text{mL sample}}$$

where A = mL phenylarsine oxide titrant (0.00564 N)

4.8 Residual Chlorine
DPD Colorimetric Method

A. General

1. Reference: *Standard Methods* (1989, pp. 4–62 through 4–64) and EPA (1983, Method 330.5).
2. Outline of Method:
   When N,N-diethyl-p-phenylene diamine (DPD) is added to a sample containing free chlorine residual, an instantaneous reaction occurs, producing a red color. The intensity of the red color, which is proportional to the residual concentration, can be determined by colorimetric analysis.

B. Special Reagents

1. *Hach DPD Free Chlorine Residual Powder Pillows*: These should be suitable for 25 mL samples.
2. *Hach DPD Total Chlorine Residual Powder Pillows*: These should be suitable for 25 mL samples.
3. *Stock Potassium Permanganate Solution*: Weigh 0.891 g of potassium permanganate ($KMnO_4$) and transfer to a 1 L volumetric flask. Dissolve the $KMnO_4$ in approximately 200 mL of DDW and then dilute to the mark with DDW.
4. *Standard Potassium Permanganate Solution (10 mg/L Chlorine Equivalent)*: Pipet 10 mL of stock $KMnO_4$ solution into a 1 L volumetric flask and dilute to the mark with DDW.

C. Preparation of Standard Series of Dilutions for Calibration Plot

Table 4.7 gives the standard chlorine equivalent concentration and the amount of working solution used in each standard. Transfer standard solutions with volumetric pipets and dilute the solutions to 100 mL in volumetric flasks using DDW.

Place a 25 mL sample of the first dilution series into a 125 mL Erlenmeyer flask. Add one DPD powder pillow (Hach). Swirl to mix. Wait 1 min for color development. Transfer to the cuvette and read and record the absorbance at 515 nm. Repeat 4.6 D and E for each standard in the series. Prepare a calibration plot from the absorbance of the standards.

Table 4.7 Chlorine Equivalent Concentrations for Calibration.

Solution Number	Equivalent Standard Concentration (mg/L)	Volume of Standard $KMnO_4$ Transferred (mL) (Diluted to 100 mL)
0	0.0	0
1	0.5	5
2	1.0	10
3	2.0	20
4	3.0	30
5	4.0	40

D. Procedure

Free Chlorine Residual

1. To a 25 mL sample in a 125 mL Erlenmeyer flask, add one DPD free chlorine residual powder pillow (Hach).
2. Swirl to mix, wait 1 min for color development, transfer to a cuvette, and read the adsorbance at 515 nm.

Total Chlorine Residual

1. The same procedure is followed as above except use a total chlorine residual (Hach) powder pillow and let stand 2 min prior to reading the absorbance at 515 nm.

E. Calculations

Calculate the concentration of free or total chlorine residual using a linear regression of the calibration standards, or plot the absorbance of the calibration standards against the chlorine equivalent concentrations and compute the sample concentration directly from the linear curve. When the chlorine residual exceeds 4 mg/L, dilute the sample with chlorine demand free DDW prior to analysis.

4.9 Cyanide
Titrimetric and Ion-Selective Electrode Methods

Part One: Sample Pretreatment

A. General

1. References: See *Standard Methods* (1989, pp. 4–20 through 4–34) and EPA (1983, Method 335.2).
2. Outline of Method:
   The complexed cyanide is converted to hydrogen cyanide gas in the presence of sulfuric acid. The gas is distilled and absorbed in a sodium hydroxide solution. Subsequent analysis is for sodium cyanide. Most interferences are eliminated in the distillation process. However, sulfide, fatty acids, and aldehydes will interfere in the distillation and analysis of cyanide. See *Standard Methods* (1989, p. 4–20) for a discussion of interfering substances.
3. Maximum holding time for cyanide is 24 hr. Preserve samples with 6 N NaOH (~2 mL per 500 mL sample).

   Note: Cyanide samples and hydrogen cyanide gas are extremely toxic. Avoid contact, ingestion, or inhalation. Process samples in a hood.

B. Special Reagents and Materials

1. *Sodium Hydroxide Solution, 1.25 N*: Dissolve 50 g of NaOH in DDW and dilute to 1 L.
2. Use *Concentrated Sulfuric Acid (H_2SO_4).*
3. *Magnesium Chloride Solution*: Weigh 510 g of $MgCl_2 \cdot 6H_2O$ into a 1000 mL flask, dissolve, and dilute to 1 L with DDW.
4. *Reflux distillation apparatus shown in Figure 4.1*: The boiling flask should be 1 L in size.
5. *Stock Standard Potassium Cyanide (1000 mg/L)*: Dissolve 2.51 g KCN and 2 g KOH in DDW. Dilute up to 1 L.

   Note: KCN is extremely toxic; avoid contact or inhalation, or contact with any acidic solutions.

6. *6 N Sodium Hydroxide*: Dissolve 240 g NaOH in DDW and dilute to 1 L.

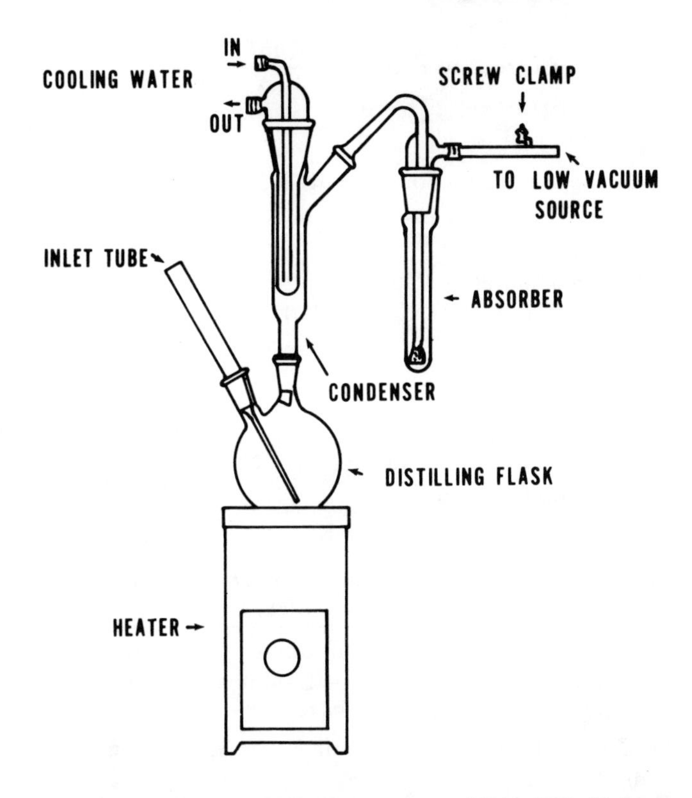

Figure 4.1 Cyanide distillation apparatus (EPA, 1983, Method 335.2).

7. *Silver Nitrate Titrant*: 0.0192 N $AgNO_3$. Dissolve 3.2617 g in DDW and dilute to 1 L.
8. *Indicator Solution for Titration*: Dissolve 20 mg p-dimethylamino-benzalrhodanine in 100 mL acetone.

C. Standardization

1. Standardize the KCN stock. Pipet 25.0 mL KCN stock into a 125 mL beaker or flask. Add 0.5 mL indicator solution and titrate with the silver nitrate titrant. 25 mL of stock should take 23–25 mL titrant. The color change is from yellow to salmon. Titrate a blank using the same sample volume and base concentration.

$$\text{Correction Factor} = \frac{25 \text{ mL Stock}}{(A - B)}$$

where

A = mL titrant for stock KCN, and
B = mL titrant for blank

Multiply each standard solution by this correction factor, e.g., (1 mg/L) (C.F.) = corrected standard.

2. Standard Solutions:

Note: To keep all cyanide solutions basic, add 0.5 mL 6 N NaOH to the 100 mL volumetrics and 5.0 mL 6 N NaOH to the 1 L volumetrics before diluting to volume.

Final Concentration mg CN^-/L	mL of Stock Solution 1000 mg/L
100	10.0 mL diluted up to 100 mL
10	10.0 mL diluted up to 1 L

Final Concentration mg CN^-/L	mL of 10 mg/L Solution
1.0	10.0 mL diluted up to 100 mL
0.1	1.0 mL diluted up to 100 mL
0.05	5.0 mL diluted up to 1 L

It is recommended that two standards be distilled, as a check on the distillation technique. For the above standards, distill the 10 mg/L and the 0.05 mg/L solutions. Follow the technique outlined in 4.7 D.

D. Procedure

1. Place 500 mL of sample or standard in the 1 L boiling flask. If the sample contains more than 500 mg CN^-, use an aliquot diluted to 500 mL. Add 5–10 boiling chips.
2. Add 50 mL of 1.25 N sodium hydroxide solution to the absorbing tube and dilute if necessary with DDW to obtain an adequate depth of liquid in the absorber. Connect the boiling flask, condenser, absorber, and trap.

3. Start a slow stream of air entering the boiling flask by adjusting the vacuum source. Adjust the vacuum so that approximately one bubble of air per second enters the boiling flask through the air inlet tube. CAUTION: The bubble rate will not remain constant after the reagents have been added and while heat is being applied to the flask. It will be necessary to readjust the air rate occasionally to prevent the solution in the boiling flask from backing up into the air inlet tube.
4. Slowly add 25 mL concentrated sulfuric acid through the air inlet tube. Rinse the tube with DDW and allow the airflow to mix the flask contents for 3 min. Pour 20 mL of the magnesium chloride solution into the air inlet and wash down with DDW.
5. Heat the solution to boiling, taking care to prevent the solution from backing up into and overflowing from the air inlet tube. Reflux for 1 hr. Turn off the heat and continue the airflow for at least 15 min. After cooling the boiling flask, disconnect the absorber and close off the vacuum source.
6. Drain the solution from the absorber into a 250 mL volumetric flask and bring up to volume with DDW washings from the absorber tube.
7. Clean the distillation apparatus well between samples.

Part Two: Analysis
Titrimetric; Ion-Selective Electrode Method

A. General

1. References: See *Standard Methods* (1989, pp. 4-30 through 4-34) and EPA (1983, Method 335.2).
2. Outline of Method:
   For the titrimetric analysis (CN^- concentration greater than 1 mg/L), a silver nitrate solution is used to titrate cyanide in the presence of a silver sensitive indicator. For the electrode method, a CN^- ion-selective electrode is used in combination with a double junction reference electrode. Use the ion-selective electrode method for the CN^- concentration range from 0.05 to 10 mg CN^-/L.

B. Special Reagents and Materials

1. *Silver Nitrate Solution and Titrimetric Indicator Solution*: See 4.9, Part One.
2. Use a *CN^- Ion-Selective Electrode, a Double Junction Reference Electrode, and a pH Meter with Expanded Millivolt Scale.*
3. *Standards*: See 4.9, Part One.
4. *Spike Solution for Ion-Selective Electrode Method*: Dissolve 0.0673 g AgCN, 0.0327 g KCN, and 1 g KOH in DDW. Dilute to 1 L.

Titrimetric

C. Standardization

See 4.9, Part One. Set up only the 10 mg/L and the 0.05 mg/L standards.

D. Procedure

1. Measure a volume of the absorption solution so that the titration will require less than 10 mL of $AgNO_3$ titrant. Dilute to a convenient volume (up to 250 mL) for titration. For samples with less than 5 mg/L cyanide concentration do not dilute. Add 0.5 mL of indicator solution.
2. Using standard $AgNO_2$ titrant, titrate to the first change in color from canary yellow to a brownish-pink.

3. Check the distillation procedure by titrating the two standards that were distilled.

E. Calculation

$$\text{mg CN}^-/\text{L} = \frac{(A - B) \times 1000}{\text{mL original sample}} \times \frac{250}{\text{mL portion used}}$$

where

A = mL of titrant used for the sample
B = mL of titrant used for the blank

Ion-Selective Method

C. Standardization

See 4.9, Part One. Set up all standards.

D. Procedure

1. Refill the double junction reference electrode with double junction outer solution and double junction inner solution in the respective compartments.
2. Connect cyanide electrode and double junction reference electrode to a digital pH/millivolt meter. Set meter scale to millivolt.
3. Use a 50.0 mL portion of each standard and each sample. Add 1.0 mL spike solution $AgK(CN)_2$ to each 50 mL portion of the standards and samples.
4. Soak electrodes in the 0.1 mg/L solution for 10 min while stirring. For all samples and standards, stir at a constant speed and keep at a constant temperature as much as possible.
5. Rinse electrodes in DDW; place electrodes in the 0.05 mg/L standard solution to a depth of 1 in. Again, stir at a constant rate and allow 10 min for a stabilized reading. Record the millivolt reading.
6. Read the samples; allow 5 min for a stabilized reading for all standards and samples. Record the millivolt readings.
7. Read standard solutions of 0.1 mg/L to 100 mg/L. Record the millivolt readings.

Table 4.8 Standard Cyanide Concentrations after Distillation.

Concentration Set Up	Concentration after Distillation	Millivolts
~10 mg/L	~20 mg/L	. . .
~0.05 mg/L	~0.1 mg/L	. . .

E. Calculation

1. *Dilution factor for distilled standards.* For each of the standards that was distilled, the CN^- concentration in the final solution was twice the initial concentration (500 mL original, 250 mL after distillation). Therefore, correct the standards in a list of concentration vs millivolts. (See Table 4.8.)
2. *Sample Concentration.* Plot millivolts vs cyanide ion concentration on semilog paper (CN^- is log axis) or use a calculator with a linear regression program. Determine the cyanide concentration of samples using this curve. Correct the sample concentration due to distillation (500 mL original, 250 mL after distillation) by multiplying the number obtained from the curve by 0.5.

F. Notes

1. Rinse the cyanide electrode thoroughly with DDW, dry, and return to the electrode box for storage.
2. Rinse reference electrode and store in DDW.
3. The $AgK(CN)_2$ spike solution is added to aid in the detection of cyanide present. It is not used in the same manner as spiking solutions defined in 1.5.

4.10 Fluoride
SPADNS Method

A. General

1. Reference: *Standard Methods* (1989, pp. 4–84 through 4–85; 4–89 through 4–91) and EPA (1983, Method 340.1).
2. Outline of Method:
   The colorimetric method depends upon the reaction between fluoride and a zirconium dye (red in color). Contrary to most photometric analyses, the more fluoride there is present, the more dye dissociates into a colorless complex ($ZrF_6^=$) and the less colored the solution becomes. The resultant standard curve has a negative slope. The addition of the highly colored SPADNS reagent must be done with utmost accuracy; a small mistake in reagent addition is the most prominent source of error in this test. See *Standard Methods* for a discussion of interferences. Some samples may require distillation to remove interferences. Store samples in polyethylene containers.
3. Use the SPADNS method for samples with a fluoride concentration of 0.05 to 1.4 mg/L.

B. Special Reagents

1. *SPADNS Solution*: Using DDW, dissolve 958 mg sodium 2-(p-sulfophenylazo)-1, 8-dihydroxy-3, 6-naphthalene disulfonate, also called 4,5-dihydroxy-3-(parasulfophenylazo)-2,7-napthalenedisulfonic acid trisodium salt, in a 500 mL volumetric flask and dilute to volume with DDW. The solution will remain stable almost indefinitely.
2. *Zirconyl-Acid Reagent*: Dissolve 133 mg zirconyl chloride octahydrate, $ZrOCl_2 \cdot 8H_2O$, in about 25 mL DDW in a 500 mL volumetric flask. Add 350 mL conc HCl and dilute to volume with DDW.
3. *Acid Zirconyl-SPADNS Reagent*: Mix equal volumes of SPADNS solution and zirconyl-acid reagent. This will remain stable 2 years.
4. *Reference Solution*: Add 10 mL of SPADNS solution to 100 mL DDW. Dilute 7 mL conc HCl to 10 mL with DDW and add to the diluted SPADNS solution. This will remain stable indefinitely.
5. *Sodium Arsenite Solution*: Dissolve 5.0 g $NaAsO_2$ in DDW in a 1 L volumetric flask and dilute to volume with DDW.

6. *Stock Fluoride Solution*: Dissolve 221.0 mg anhydrous sodium fluoride, NaF, in DDW and dilute to 1 L.

$$1.00 \text{ mL} = 100 \ \mu F^-$$

C. Standardization

1. *Standard Fluoride Solution*: Dilute 10.0 mL stock fluoride standard to 100 mL with DDW.

$$1.00 \text{ mL} = 0.01 \text{ mg } F^-; \ 10 \text{ mg}F^-/L \text{ standard}$$

2. Using the 10 mg/L standard, set up the following dilutions for a standard curve.

Final Concentration mg/L	mL of the 10 mg/L Standard
0.1	1.0 mL diluted up to 100 mL
0.5	5.0 mL diluted up to 100 mL
1.0	10.0 mL diluted up to 100 mL
1.4	7.0 mL diluted up to 50 mL

Treat the standards the same as the samples in the following procedure.

3. The absorbance values to be expected for the blank minus the standard will vary somewhat as batches of reagents are prepared. Check with the laboratory supervisor or recent data books to determine absorbance ranges for the standards and blanks.

D. Procedure

1. If the sample contains residual chlorine, eliminate it by adding 1 drop of the arsenite solution for each 0.1 mg Cl_2 and mix.
2. Using a 25 mL sample or standard solution, or an aliquot diluted to 25 mL, add 5.0 mL of the acid-zirconyl-SPADNS reagent. *Mix well.* Be sure that the sample temperature is the same as that of the standards.
3. Zero the spectrophotometer using the reference solution instead of DDW. Read the samples and standards at 570 nm, using a 1 cm cell.

   Note: Remember that this is an inverse curve. The blank will have the highest absorbance reading at 570 nm.

E. Calculation

Calculate the concentration of F^- in a sample using a linear regression of the calibration standards, or plot the absorbance of the calibration standards against the calibration concentrations and compute the sample concentration directly from the linear standard curve. If the sample was diluted, multiply by the appropriate dilution factor.

4.11 Low-Level Fluoride Ion-Selective Electrode Method

A. General

1. References: See *Standard Methods* (1989, pp. 4–84 through 4–85; 4–87 through 4–89) and EPA (1983, Method 340.2).
2. Outline of Method:
   The fluoride ion-selective electrode can be used to measure the activity or the concentration of fluoride in aqueous samples by use of an appropriate calibration curve. However, the fluoride activity depends on the total ionic strength of the sample. The electrode does not respond to bound or complexed fluoride. A buffer solution of high ionic strength may be added to swamp variations in sample ionic strength.

B. Special Reagents and Equipment

1. *pH Meter*: Use an expanded scale or digital (or ion-selective meter).
2. *Electrode, Reference*: Use a sleeve-type Orion #90–01–00, Beckman #90463, Corning #47612, or the equivalent.
3. *Electrode, Fluoride*: Use Orion #94–09 or the equivalent.
4. Use a *Magnetic Stirrer*.
5. *Total Ionic Strength Adjustment Buffer (TISAB)*: To approximately 500 mL DDW in a 1 L beaker add 57 mL glacial acetic acid, 58 g NaCl, and 4.0 g 1, 2 cyclohexylenediaminetetraacetic acid (CDTA). (Low-level samples containing less than 0.4 mg/L fluoride and containing no complexing agents such as aluminum or iron may be run using TISAB without CDTA.) Stir to dissolve. Place beaker in a cool water bath for cooling. Immerse a calibrated pH electrode into the solution and slowly add approximately 125mL 6N NaOH until the pH is between 5.0 and 5.5. Cool to room temperature and dilute to 1 L.
6. *Stock Fluoride Solution; 100 mg F^-/L*: Dissolve 221.0 mg anhydrous sodium fluoride, NaF, in DDW and dilute to 1 L.

$$1.00 \text{ mL} = 100 \ \mu\text{g } F^-$$

Table 4.9 Increments of Fluoride Solution (10 mg/L) Added for Developing the Standard Curve.

Step	Added Volume (mL)	Concentration (mg/L)
1	0.1	0.02
2	0.1	0.04
3	0.2	0.08
4	0.2	0.12
5	0.4	0.20
6	2.0	0.57
7	2.0	0.91
8	2.0	1.23
9	2.0	1.53
10	2.0	1.80
11	2.0	2.06
12	2.0	2.31

C. Standardization

1. Prepare a dilute standard solution by adding 10.0 mL of stock fluoride solution to DDW and diluting to 100 mL.

$$1.00\ \text{mL} = 10\ \mu\text{g}\ F^-;\ 10\ \text{mg}\ F^-/\text{L}$$

2. Calibration curve for low-level measurements:
   a. To a 150 mL plastic beaker add 50 mL TISAB and 50 mL DDW. Set the function switch to relative millivolts (REL MV). Place electrodes in solution and stir thoroughly.
   b. Add increments of the 10 mg/L F^- standard using the steps in Table 4.9. Record the electrode potential in millivolts (mv) after each increment. Make certain the readings are stable.

D. Procedure

1. Add 50 mL TISAB to a 50 mL sample.
2. Rinse electrodes, blot dry, and place in the sample. Stir thoroughly.
3. Record the electrode potential for the sample when the reading is stable.

E. Calculations

1. Using a calculator with linear regression capabilities, plot the log F^- concentration (mg/L) vs the electrode potential (mv) to generate a calibration curve. Determine the concentration of the sample from the calibration curve.
2. As an alternate method, use semilogarithmic graph paper to plot the standard concentrations (log axis) against the potential (linear axis) to determine the calibration curve. Determine the concentration of the sample from the calibration curve.

F. Notes

Samples and standards should always be stored in plastic containers as fluoride reacts with glass.

4.12 Ammonia
Preliminary Distillation Step
Nesslerization and Titrimetric Methods

A. General

1. References: *Standard Methods* (1989, pp. 4–111 through 4–120; 4–121 through 4–122) and EPA (1983, Method 350.2).
2. Outline of Method:
   The sample is buffered at pH 9.5 with a borate buffer. It is distilled into a solution of boric acid if Nesslerization or titration techniques are to be used and into sulfuric acid if the indophenol technique is to be used. The technique selected depends on the concentration of ammonia in the sample.

mg NH_3-N/L	Technique
0.010–0.500	Low-level indophenol
0.500–5.00	Nesslerization or high-level indophenol
>5.00	Titration or high-level indophenol

   Also, when using the indophenol technique, distill the sample if the alkalinity is greater than 500 mg/L or if the sample is turbid.

B. Special Reagents

1. *Ammonia-Free Water*: Ordinary distilled water is generally not ammonia-free. Use fresh DDW from the purification system at all times.
2. *Sodium Hydroxide, 6 N*: Dissolve 240 g NaOH in fresh ammonia-free DDW. Cool and dilute to 1 L.
3. *Sodium Hydroxide, 0.1 N*: Dissolve 4 g NaOH in fresh ammonia-free DDW and dilute to 1 L.
4. *Borate Buffer Solution*: Dissolve 4.75 g sodium tetraborate ($Na_2B_4O_7 \cdot 10H_2O$) in fresh ammonia-free DDW and dilute to 500 mL. Transfer quantitatively to a 1 L volumetric flask. Add 88 mL 0.1 N NaOH solution and dilute to 1 L with fresh ammonia-free DDW.

5. *Dechlorinating Agent N/70, Sodium Thiosulfate*: Dissolve 3.5 g $Na_2S_2O_3 \cdot 5H_2O$ in fresh ammonia-free DDW and dilute to 1 L. Prepare fresh weekly. Use 1 mL to remove 1 mg/L residual chlorine in a 500 mL sample.
6. *Neutralization Agents*:
   a. *Sodium Hydroxide, 1 N*: Dissolve 40 g NaOH in fresh ammonia-free DDW and dilute to 1 L.
   b. *Sulfuric Acid, 1 N*: Add 28 mL conc H_2SO_4 carefully to 500 mL fresh ammonia-free DDW. Cool and dilute to 1 L.
7. *Absorbent Solution, Boric Acid*: Dissolve 20 g H_3BO_3 in fresh ammonia-free DDW and dilute to 1 L. Use with the Nesslerization technique.
8. *Absorbent Solution, Sulfuric Acid*: Add 1.0 mL conc H_2SO_4 to fresh ammonia-free DDW and dilute to 1 L. Use with the indophenol technique.
9. *Absorbent Solution, Indicating Boric Acid*: Use with the titration technique.
   a. Mixed Indicator: Dissolve 200 mg methyl red indicator in 100 mL 95% ethyl alcohol. Dissolve 100 mg methylene blue in 50 mL 95% ethyl alcohol. Combine the two solutions and mix well.
   b. *Absorbent Solution*: Dissolve 20 g H_3BO_3 in fresh ammonia-free DDW, add 10 mL mixed indicator and dilute to 1 L. Both a and b are only stable for about 1 month.
10. *Titration Reagents*:
    Standard H_2SO_4, 0.02 N Titrant: To standardize, use the technique described in 4.2. For the greatest accuracy, standardize the titrant against an amount of Na_2CO_3 that has been added to the indicating boric acid solution:

$$1.0 \text{ mL } H_2SO_4(0.0200 \text{ N}) = 280 \ \mu\text{g N}$$

11. *Nessler Reagent*: Dissolve 100 g mercuric iodide (HgI_2) and 70 g potassium iodide (KI) in a small quantity of fresh ammonia-free DDW in a beaker. In a 1 L volumetric flask, dissolve 160 g sodium hydroxide (NaOH) in 500 mL fresh ammonia-free DDW with stirring. Cool thoroughly and *slowly* add the HgI_2-KI solution to the volumetric flask with stirring. Dilute to 1 L. Store in a stoppered Pyrex container out of direct sunlight.
12. *Stock Standard Ammonium Chloride*: Dry 2–3 g of reagent grade ammonium chloride (NH_4Cl) at 103°C for 1 hr. Cool in a desiccator.

Dissolve exactly 1.91 g NH_4Cl in about 400 mL fresh DDW and dilute to exactly 500 mL. This solution will remain stable for about six months.

13. *Glassware Wash, 6 N HCl*: Add 500 mL concentrated HCl to 500 mL DDW in a hood. Mix and allow to cool. Reuse but discard when the solution turns yellow.
14. *Distillation Flasks*: See 4.18.

C. Standardization

Prepare one of the following standards in duplicate:

Standard I: Use when the NH_3-N concentrations in the samples are > 5 mg/L. Dilute 2.0 mL of the stock NH_4Cl standard to 200 mL with fresh ammonia-free DDW. Solution is 10 mg NH_3-N/L.

Standard II: Use when the NH_3-N concentrations in the samples are in the range 0.5 to 5 mg/L. Dilute 1.0 mL of the stock NH_4Cl standard to 500 mL with fresh ammonia-free DDW. Solution is 2 mg NH_3-N/L.

Standard III: Use when the NH_3-N concentrations in the samples are less than 0.5 mg/L. Dilute 1.0 mL of the stock NH_4Cl standard to 100 mL with fresh ammonia-free DDW. This solution is 10 mg NH_3-N/L. Dilute 4.0 mL of this solution to 100 mL with fresh ammonia-free DDW. Solution is 200 μg NH_3N/L.

Carry the appropriate standard and blanks in duplicate through the procedure below.

D. Procedure

Note: Wash glassware with 6 N HCl and rinse 4–5 times with DDW.

1. Add 500 mL fresh ammonia-free DDW and 20 mL borate buffer to a Kjeldahl distillation flask and adjust pH to 9.5 with 6 N NaOH solution. Add a few boiling chips and use this solution to steam out the distillation apparatus.
2. Place a 500 mL sample (or an aliquot diluted to 500 mL with fresh ammonia-free DDW) and boiling chips in a Kjeldahl flask. If the

ammonia concentration in a sample is less than 100 $\mu g/L$, use a sample volume of 700 mL. (A larger sample volume may be used if distillation flasks larger than 800 mL are used). If necessary, dechlorinate with sodium thiosulfate. Adjust the pH to approximately 7 with 1 N sodium hydroxide or sulfuric acid as necessary.

3. Add 25 mL borate buffer solution and adjust to pH 9.5 with 6 N NaOH.
4. Distill sample immediately after steaming out distillation apparatus. Distill at a rate of 6 to 10 mL/min.
5. Collect sample distillate with the tip of the delivery tube submerged in 50 mL of absorbent in a 500 mL flask.

   a. Absorbent

      i. Indophenol: 0.04 N H_2SO_4
      ii. Nesslerization: 2% H_3BO_3
      iii. Titration: 2% indicating H_3BO_3

   b. If H_3BO_3 is used, add additional 50 mL increments of H_3BO_3 for each mg NH_3-N distilled.
6. Collect about 300 mL of distillate. Lower the delivery flask (free of contact with the delivery tube) during the last minute or two of distillation and allow the condenser and delivery tube to be cleansed. Dilute sample to 500 mL with fresh ammonia-free DDW using a volumetric flask.
7. Continue distillation so as to concentrate residue in Kjeldahl flask to 40–50 mL—throw excess distillate away. Save the residue for organic nitrogen determination, if desired.
8. Test for NH_3-N in the distillate by either titration, Nesslerization, or indophenol techniques.

F. Notes

1. Preparation of glassware and distilling apparatus is of EXTREME IMPORTANCE for maintaining a high level of accuracy. Glassware used in this test should be set aside and marked for AMMONIA ANALYSIS ONLY. After washing glassware, the distillation apparatus should be steamed.
2. Acid titration
   Titrate distillate with standard 0.02 N H_2SO_4 titrant until the indicator turns a pale lavender in color.

$$\text{mg NH}_3\text{-N/L} = \frac{(D - E) \times 280}{\text{mL sample}}$$

D: mL H_2SO_4 for sample
E: mL H_2SO_4 for blank

3. Nesslerization
   a. Measure a 50 mL aliquot of the distillate into a clean beaker that has been carefully rinsed with ammonia-free DDW.
   b. Add exactly 2 mL of Nessler's reagent to the sample with a safety pipeting bulb. Draw the Nessler's from near the surface using extreme caution not to disturb the precipitate that settles to the bottom of the reagent bottle.
   c. Mix the Nesslerized sample immediately; wait exactly 20 min. Mix again and read the absorbance at 410 nm.
4. Indophenol Technique
   Before analysis, adjust the sample solution pH to approximately 7 using 6 N NaOH. Then analyze this according to the indophenol technique in 4.13.

4.13 Ammonia
Low-Level Indophenol Method

A. General

1. References: See Solorzano (1969, pp. 799–801) and *Standard Methods* (1989, pp. 4–111 through 4–114; 4–120 through 4–121).
2. Outline of Method:
   The blue color of indophenol obtained by the reaction of ammonia, phenol, and hypochlorite at high pH is being measured. This method eliminates interference due to precipitation by complexing Mg and Ca with citrate.

B. Special Reagents

1. *Fresh doubly deionized water (DDW)*: Always use fresh DDW taken from the reagent-grade purification system immediately prior to analysis.
2. *Phenol-alcohol solution*: 10 g of phenol in 100 mL of 95% ethyl alcohol.
3. *Sodium Nitroprusside*: (Sodium nitroferricyanide, $Na_2Fe(CN)_5NO \cdot 2H_2O$). Dissolve 0.5 g in 100 mL of fresh DDW; store in a dark bottle.
4. *Alkaline solution*: 100 g of trisodium citrate and 5 g of NaOH in 500 mL fresh DDW.
5. *Sodium hypochlorite solution*: Commercial product Clorox.
6. *Oxidizing solution*: 100 mL of the alkaline solution and 25 mL of the hypochlorite solution mixed together and used the same day of analysis.
7. *Hydrochloric Acid, 6 N*: (Glassware wash). Carefully add 500 mL conc HCl to 500 mL DDW. Cool. Reuse and discard when the solution turns yellow.
8. *Stock Ammonium Chloride Standard Solution*: Dry 2–3 g of reagent grade NH_4Cl at 103°C for 1 hr. Cool in a desiccator. Dissolve 1.91 g NH_4Cl in 490 mL fresh DDW and dilute to 500 mL.

$$1.00 \text{ mL} = 1 \text{ mg } NH_3\text{-N}$$

C. Standardization

1. Standard Curve
   Range of ammonia concentrations is 10 μg/L–400 μg/L. Below 10 μg/L the absorbance is not appreciably different from the blank value. The analysis is more reliable at concentrations close to the standards within the range. Set up standards close to the concentration of the samples.
2. Procedure for daily standardization: *Always use fresh DDW.*
   a. Dilute 1.0 mL of the 1 mg/mL stock solution up to 100 mL: the solution is 10 mg/L NH_3-N.
   b. Prepare one of the following:

Final Concentration	mL of the 10 mg/L Solution
100 μg/L	2.0 mL diluted up to 200 mL
200 μg/L	4.0 mL diluted up to 200 mL

3. Set up duplicates for the standard (100 or 200 μg/L) and set up two fresh DDW blanks. Treat the standards and blanks exactly like the samples in the procedure below.
4. With a 1 cm cell, the following gives a guideline for the standards:

Standard	Range of Absorbances (std-blank) at 640 nm
100 μg/L	0.107–0.137

D. Procedure

Sample concentrations must range from 10 μg/L–400 μg/L. If the sample has a higher concentration, it must be diluted to within this range.

1. Rinse glassware with 6 N HCl and several times with fresh DDW. If possible, soak glassware overnight in 6 N HCl.
2. Add consecutively to a 50 mL filtered sample (or an aliquot diluted to 50 mL): 2 mL of phenol solution, 2 mL nitroprusside solution, and 5 mL of the oxidizing solution. *Mix well* after each addition of reagents to produce the appropriate color. Stopper the flasks well and allow the color to develop for 2 hr in the dark; the color is good for 24 hr. Do not use rubber stoppers. Use plastic or foil to cover the flasks.
3. Read and record absorbance at 640 nm using a 1 cm cell.

E. Calculation

Calculate the ammonia in the sample as follows:

$$\mu g\ NH_3\text{-}N/L = (Abs_{sample} - Abs_{blank})(m^{-1})\ (df)$$

where

$$m^{-1} = \frac{\Delta concentration}{\Delta absorbance} = \frac{100 - 0}{Abs_{std} - Abs_{blank}} = \frac{200 - 0}{Abs_{std} - Abs_{blank}}$$

df = dilution factor

If a 10% sample is used, df = 10; if a 5% sample is used, df = 20.

The concentration of the ammonia in a sample may also be calculated using a linear regression of the calibration standards; or plot the absorbance of the calibration standards against the calibration concentrations and compute the sample concentration directly from the linear standard curve.

4.14 Ammonia
High-Level Indophenol Method

A. General

1. References: See Zadorojny et al. (1973, pp. 905–912), Solorzano (1969, pp. 799–801), and *Standard Methods* (1989, pp. 4–111 through 4–114; 4–120 through 4–121).
2. Outline of Method:
   This method varies from the low-level indophenol technique (4.13) in the detection of sodium nitroprusside, which is a color intensifier. If the color is read exactly 90 min after the addition of reagents, a standard curve up to 50 mg NH_3-N/L is linear. In this procedure, the color continues to intensify with time. After 90 min the higher standards begin to lose the linearity. After 24 hr, a standard curve up to 20 mg NH_3-N/L is still linear.

B. Special Reagents

For all reagents, see 4.13. Omit the preparation and use of sodium nitroprusside (sodium nitroferricyanide).

C. Standardization

Range: 1 to 50 mg NH_3-N/L (read at 90 min)
1 to 20 mg NH_3-N/L (read after 90 min and before 24 hr)

Note: Do not use the high-level procedure for samples with a concentration less than 1 mg NH_3-N/L. Set up a series of standards close to the concentration of the samples. Below is a guideline for standard preparation. Always pour standard solutions in duplicate for analysis.

Final Concentration mg NH_3-N/L	mL Stock Standard (1000 mg/L)
5	1.0 mL diluted up to 200 mL
10	2.0 mL diluted up to 200 mL
20	4.0 mL diluted up to 200 mL
30	6.0 mL diluted up to 200 mL
40	8.0 mL diluted up to 200 mL
50	10.0 mL diluted up to 200 mL

Final Concentration mg NH_3-N/L	mL of the 10 mg NH_3-N/L Solution
1	20 mL diluted up to 200 mL
2	20 mL diluted up to 100 mL

Set up two DDW blanks. For all standards and blanks, follow the procedure below for reagent addition and color measurements.

D. Procedure

1. Rinse 125 mL Erlenmeyer flasks with 6 N HCl and 4–5 times with fresh DDW.
2. To a 50 mL sample or standard, add 2 mL of phenol solution and *mix well* by swirling. Add 5 mL of the oxidizing solution and *mix well* again. Since the color will continue to develop in the samples overtime, add reagents to the samples at some interval (1 min) to allow spectrophotometric reading at the correct time.
3. Store samples tightly covered and in the dark.

   Note: To cover the flasks, use tight-fitting aluminum foil or plastic stoppers. Do not use rubber stoppers as this may result in contamination.

4. Read samples using a 1 cm cell at 640 nm. For a standard curve up to 50 mg NH_3-N/L, read after *exactly 90 min*. For a standard curve up to 20 mg NH_3-N/L, samples can be read at a later time. But *read all samples and standards at the same time, relative to the reagent addition.*

E. Calculation

Calculate the concentration of NH_3-N in a sample using a linear regression of the calibration standards; or plot the absorbance of the calibration standards against the calibration concentrations and compute the sample concentration directly from the linear standard curve.

4.15 Nitrite-Manual Diazotization Method

A. General

1. References: See Strickland and Parsons (1972, pp. 77–80), Parsons et al. (1984, pp. 7–9), *Standard Methods* (1989, pp. 4–128 through 4–131), and EPA (1983, Method 354.1).
2. Outline of Method:
Nitrite reacts with sulfanilamide in an acid solution. The resulting diazo compound reacts with the naphthylethylenediamine reagent and forms a highly colored azo dye, the absorbance of which is measured spectrophotometrically. Applicable range is 1 μg to 200 μg NO_2^--N/L.

B. Special Reagents

1. *Sulfanilamide Solution*: Dissolve 5 g of sulfanilamide in a mixture of 50 mL of concentrated hydrochloric acid and approximately 300 mL of DDW. Dilute to 500 mL with DDW. This will remain stable many months.
2. *N-(1-Naphthyl)-Ethylenediamine-Dihydrochloride Solution*: Dissolve 0.50 g of the reagent powder in 500 mL of DDW and store in a dark bottle. This will remain stable about 1 month.
3. *Standard Nitrite Stock Solution*: A small amount of anhydrous analytical reagent quality sodium nitrite ($NaNO_2$) should be dried at 110°C for 1 hr. Dissolve 0.345 g of the dried $NaNO_2$ in DDW and dilute to 1 L. Store in dark bottle to which 1 mL of chloroform has been added as a preservative and bacterial inhibitor. (Note: When using chloroform, always work under a hood. Use safety glasses or a mask and wear gloves.) This will remain stable 1–2 months.

   Note: Standard concentration is expressed as nitrogen. See *Standard Methods* (1989) for a more complete description for the preparation of a very accurate nitrite-nitrogen standard.

$$1.00 \text{ mL} = 70\ \mu\text{g } NO_2^-\text{-N}$$

C. Standardization

1. Standard curve: 0 to 200 μg NO_2^--N/L
2. Procedure for daily standardization:
   a. Dilute 1.0 mL of the 70 μg/mL stock standard up to 100 mL; this solution is 700 μg/L.
   b. Dilute 5.0 mL of the 700 μg/L solution up to 100 mL; this solution is 35 μg/L.

3. Pour 50 mL duplicates of the 35 μg/L standard. Also, pour two 50 mL DDW blanks. Treat standards and blanks exactly like the samples in the procedure below.

D. Procedure

1. Analysis is to be performed upon GF/C filtered samples. If apparent coloration remains in the sample, run a turbidity correction test. To a 30 mL aliquot of sample, add 1 mL of sulphanilamide only and measure absorbance.
2. Measure a 50 mL sample or an aliquot diluted to 50 mL into a 125 mL Erlenmeyer flask. Make sure samples are at room temperature (15–25°C).
3. Add 1 mL of sulphanilamide solution, mix, and allow a reaction time of 2–8 min.
4. Add 1 mL of the naphthylethylenediamine solution. Complete color development requires 10 min with the color remaining stable for 2 hr.
5. Measure absorbances of the solutions following the 10 min reaction period in 1 or 5 cm cells, against DDW at 543 nm. Subtract turbidity absorbance, if necessary, to give corrected absorbance for use in the calculations below.

E. Calculation

Calculate the nitrite in the sample as follows:

$$\mu g\ NO_2^- \text{-N/L} = (Abs_{sample} - Abs_{blank})(m^{-1})(df)$$

where

$$m^{-1} = \frac{\Delta concentration}{\Delta absorbance} = \frac{35 - 0}{Abs_{std} - Abs_{blank}}$$

df = dilution factor

If a 10% sample is used, df = 10; if a 5% sample is used, df = 20.

F. Notes

Samples should be stored at 4°C and analyzed within 48 hr (EPA, 1983) to minimize bacterial conversion of NO_2^--to NO_3^-- or NH_3. *Standard Methods* (1989) cautions against acid preservation of samples to be analyzed for nitrite-nitrogen.

4.16 Nitrate-Manual Cadmium-Reduction Method

A. General

1. References: Strickland and Parsons (1972, pp. 77–80), Parsons et al. (1984, pp. 7–9), *Standard Methods* (1989, pp. 4–131 through 4–132; pp. 4–135 through 4–137), and EPA (1983, Method 353.3).
2. Outline of Method:
   The nitrate in the sample is reduced almost quantitatively to nitrite when it is run through a column containing cadmium filings loosely coated with metallic copper. The nitrite thus produced is measured by the method outlined for reactive nitrite. A correction may be made for any nitrite initially present in the sample by subtracting $\mu g\ NO_2^-N/L$ from $\mu g\ (NO_3^--N + NO_2^--N)/L$.

B. Special Reagents

1. *Concentrated Ammonium Chloride-EDTA Solution*: Dissolve 13 g of analytical quality NH_4Cl and 1.7 g disodium ethylenediaminetetraacetate (EDTA) in 900 mL DDW. Adjust pH to 8.5 with conc NH_4OH and dilute to 1 L.
2. *Dilute Ammonium Chloride-EDTA Solution*: Dilute 300 mL of conc NH_4Cl-EDTA solution to 500 mL with DDW. Store in glass or plastic.
3. *Hydrochloric Acid, 6 N*: Carefully add 500 mL conc HCl to 500 mL DDW. Cool.
4. *2% V/V Copper Sulfate*: Dissolve 10 g copper sulfate pentahydrate ($CuSO_4 \cdot 5H_2O$) in 500 mL DDW.
5. *Sulfanilamide Solution:* See 4.15.
6. *N-(1-Naphthyl)-Ethylenediamine Dihydrochloride Solution*: See 4.15.
7. *Stock Standard Concentrated Potassium Nitrate*: Dry approximately 5 g of anhydrous KNO_3 in the oven at 100°C for 1 hr. Cool and dissolve exactly 3.61 g of KNO_3 in DDW and dilute to 500 mL with DDW. This will remain stable for about 6 months.

$$1.00\ mL = 1\ mg\ NO_3^--N$$

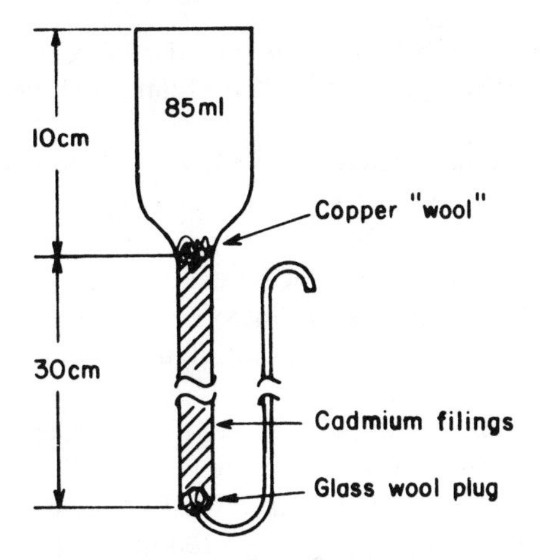

Figure 4.2 Packed reduction column.

C. Standardization

1. Standard Curve: 10 to 200 μg NO_3^--N/L.
2. Procedure for daily standardization
   a. Dilute 1.0 mL of the 1 mg/mL stock standard up to 100 mL; solution is 10 mg NO_3^--N/L.
   b. Dilute 10.0 mL of the 10 mg/L solution up to 1000 mL; this standard solution is 100 μg NO_3^--N/L.
3. Using the 100 μg/L solution as the standard, treat it exactly like the samples in the procedure below.

D. Procedure

Part One: Preparation of Reducing Columns (Figure 4.2)

Cadmium (Copper Plate) Filings

$$0.5 \text{ mm} < \text{size} < 2 \text{ mm} \sim 50 \text{ g/column}$$

1. Weigh 50 g of dried granules of cadmium for each column or empty used cadmium from the old columns.

2. Wash granules briefly with about 400 mL of 6 N HCl and then rinse 2 or 3 times with DDW. The supernatant should no longer be acidic (pH < 5).
3. Stir 100 g cadmium granules with 500 mL of 2% $CuSO_4 \cdot 5H_2O$ until all the blue color has left and semicolloidal copper particles begin to enter the supernatant liquid. Wash the copper-cadmium particles with sufficient DDW to remove any precipitated copper.
4. Plug the bottom of the reduction column with glass wool.
5. Fill the column with supernatant liquor from preparation of the cadmium.
6. Pour in the cadmium-copper granules until the depth is 30 cm. Tap the side of the column to insure the filings are *well* settled.
7. Wash the column with dilute NH_4Cl-EDTA solution.
8. Monitor the flow rate. It should be $\frac{100 \text{ mL}}{5 \text{ min}}$ to $\frac{100 \text{ mL}}{12 \text{ min}}$
9. Plug the top of the column with copper "wool." Prime each column by running 500 mL of 100 μg/L standard solution through at a rate of 7 to 10 mL/min.
10. When the column is not in use, the cadmium must be completely covered by dilute NH_4Cl-EDTA solution at all times, and the column covered with aluminum foil.
11. When the standards show that the efficiency of the column is low ($\leq$75% recovery), the cadmium filings need to be retreated.
    a. Empty the cadmium from four columns into a beaker and stir vigorously with 300 mL of 6 N HCl solution.
    b. Decant the acid and repeat the procedure.
    c. Wash the metal with 200–300 mL portions of DDW until the supernatant is no longer acidic (pH > 5); decant the liquid to leave the metal as dry as possible.
    d. Retreat the metal with $CuSO_4$ solution (see 4.16 D, step 3). There should be enough regenerated cadmium to make three columns.

Part Two: Technique for Nitrate Analysis

1. Calibrate the rate of flow in the reducing columns with dilute NH_4Cl-EDTA so that 10 mL of solution passes through the column in 1 min. Allow the dilute NH_4Cl-EDTA solution used in the calibration to drain out.
2. Measure 100 mL of GF/C or 0.45 μm filtered sample into a 250 mL beaker. The concentration of the samples should not be higher than 200 μg NO_3-N/L. Samples known to exceed this concentration should be diluted prior to reduction. Add 2 mL of conc NH_4Cl-EDTA to each of

the 100 mL samples and mix. The slight acidification of the sample by the addition of NH_4Cl-EDTA greatly slows the deactivation process; the column is good for at least 100 samples.

3. Pour in 5 mL of sample into the column and allow it to pass through. Add the remaining sample into the rinsed, drained columns and collect the first 40 mL of sample in a 50 mL graduated cylinder. Discard this portion of the sample and collect the remaining portion in the 250 mL beaker. The passage of the 40 mL flushes the column of the preceding sample.
4. As soon as the sample has passed through the column, measure a 50 mL aliquot in a 50 mL graduated cylinder, discard the remaining portion, and return the 50 mL aliquot to the beaker.
5. If the above procedure is followed, it is not necessary to rinse either the graduated cylinder or the columns between samples unless extreme changes from high to low concentrations are encountered.
6. Following the method for reactive NO_2^--N determination, obtain the absorbance of the reduced sample at 543 nm in a 1 or 5 cm cell.

E. Calculation

1. Calculate the nitrate-plus nitrite-nitrogen in the sample as follows:

$$\mu g(NO_3^- + NO_2^-) - N/L = (Abs_{sample} - Abs_{blank})(m^{-1})(df)$$

where

$$m^{-1} = \frac{\Delta concentration}{\Delta absorbance} = \frac{100 - 0}{Abs_{std} - Abs_{blank}}$$

$$df = \text{dilution factor}$$

If a 10% sample is used, df = 10; if a 5% sample is used, df = 20.

2. Using the value from the above calculation, nitrate-nitrogen can be calculated by subtracting the nitrite-nitrogen from the total:

$$\mu g\ NO_3^- - N/L = \mu g(NO_3^- + NO_2^-)\text{-}N/L - \mu g(NO_2^-\text{-}N)/L$$

Note 1: For determination of NO_2^--N, refer to 4.15.
Note 2: All the units must be the same in the above equation (i.e., μg/L).

F. Notes

Samples follow Beer's law and may be diluted after the reagents have been added up to a 1→2 dilution. Samples that require more than a 1→2 dilution should be diluted *before reduction*.

4.17 Nitrate; Nitrite
Colorimetric/Automated Cadmium-Reduction Methods

A. General

1. References: See *Standard Methods* (1989, pp. 4–131 through 4–132; 4–137 through 4–139), EPA (1983, Method 353.2), Technicon Industrial Systems (1978, Industrial Method No. 102–70W/B).
2. Outline of Method:

   Nitrate-nitrogen (NO_3^--N) is reduced to nitrite-nitrogen (NO_2^--N) by contact with copper-coated cadmium granules packed in a small column. The NO_2^--N produced is diazotized with sulfanilamide under acidic conditions and then coupled with N-(1-naphthyl)-ethylenediamine dihydrochloride to form a reddish-purple azo dye. The dye is determined spectrophotometrically using a continuous colorimeter. NO_2^--N is analyzed concurrently without the use of the reduction column.

B. Special Reagents and Equipment

1. *Chloroform, $CHCl_3$*: Store in a glass-stoppered dropper bottle. This is used for sample preservation. (Note: When using chloroform, always work under a hood. Use safety glasses or a mask and wear gloves.)
2. Use *Technicon AutoAnalyzer II, or Equivalent.*

   Note: Consult above references for directions to make reagents and columns.

C. Standardization

1. 1.0 mg/L NO_3^--N and 50 μg/L NO_2^--N standards are run daily by the AutoAnalyzer technician to calibrate the instrument. The detection limit for nitrate plus nitrite is 0.04 mg/L and the upper limit of the range is 2.00 mg/L. Nitrite is run within the range of 2 to 100 μg/L.

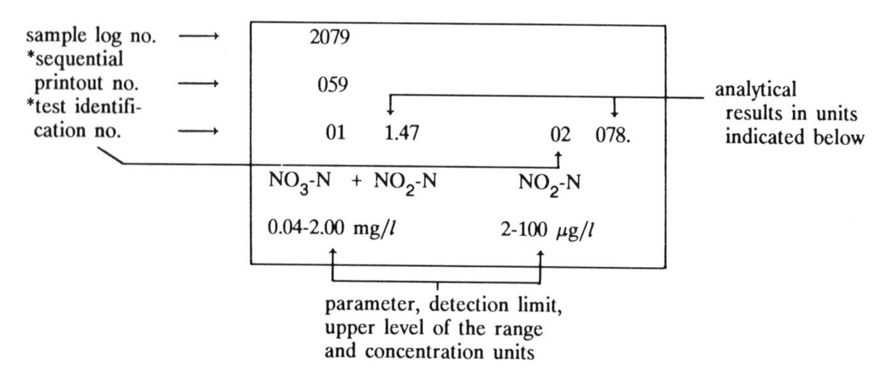

*These are for use by the AutoAnalyzer technician.

Figure 4.3 A representation of the AutoAnalyzer printout.

D. Procedure

Note: The AutoAnalyzer and similar automated analyzers require considerable expertise to run them well and a great deal of familiarity to handle troubleshooting. Follow the directions to prepare samples.

1. Fill an acid-washed (6 N HCl) and rinsed 12 mL screw cap vial with filtered sample. If replicate determinations are desired, fill more than one vial. Label with the log number of the sample using an adhesive cloth label.
2. Place one drop of chloroform in each vial as a preservative and biological inhibitor. Replace the cap and invert to mix.

   Note 1: It is very important not to use an excessive amount of chloroform. Large amounts will dissolve the plastic sample cups on the analyzer and clog the sample and reagent lines.

   Note 2: There are other preservatives available if desired (see above references).

3. Place sample vials in numerical order in a rack and label with a name and project number. Store at 4°C in the space provided for AutoAnalyzer samples.

   Note: If an occasional dilution is necessary, this will be performed by the AutoAnalyzer technician. However, if it is known that the samples will read higher than the ranges listed above, then do appropriate dilutions before filling the vials. Some samples may be very high in one parameter and low in the other. In this case, it may be necessary to analyze both a straight and a diluted sample to get reliable values for both parameters. As an aid to the AutoAnalyzer technician, label straight and diluted samples clearly.

E. Calculation

Figure 4.3 is a representation of the AutoAnalyzer printout and will aid in its interpretation. The instrument converts sample peaks directly to concentration, so calculations are unnecessary. However, to get a value for NO_3^--N alone, you must subtract NO_2^--N from the total:

$$\text{mg } NO_3^- - N/L = \text{mg } (NO_3^- - N + NO_2^- - N)/L - \frac{\mu g NO_2^- - N/L}{1000}$$

For the data above, nitrate = 1.39 mg NO_3^--N/L.

If a sample requires a dilution, the printout will show a number somewhat higher than the upper limit of the range. Look for the diluted and rerun sample later on the printout sheet.

Often the instrument will detect numbers less than the detection limit and print them. These values are not reliable and the data should be reported as less than the detection limit (i.e., <0.04 mg/L).

4.18 TKN
Total Kjeldahl and Organic Nitrogen
Digestion, Distillation, and Nesslerization Methods

A. General

1. Reference: *Standard Methods* (1989, pp. 4–143 through 4–149).
2. Outline of Method:
   In the presence of a strong acid and catalysts, organically bound nitrogen is quantitatively converted to ammonia, which is distilled (to remove impurities) and collected in boric acid. Ammonia is then determined by a Nesslerization spectrophotometric technique.

B. Special Reagents and Equipment

1. *Doubly Deionized Water (DDW)*: Always use fresh DDW from the reagent grade purification system. Alternately, use water prepared by the passage of distilled water through an ion exchange column containing a strongly acidic cation exchange resin mixed with a strongly basic anion exchange resin. Regeneration of the column should be carried out according to the manufacturer's instructions.

   Note: All solutions must be made with ammonia-free (fresh) DDW.
2. *Mercuric Sulfate Solution*: Dissolve 8 g red, mercuric oxide (HgO) in 50 mL of 1 + 4 sulfuric acid/DDW (10.0 mL conc H_2SO_4 plus 40 mL fresh DDW) and dilute to 100 mL with fresh DDW.
3. *Sulfuric Acid-Mercuric Sulfate-Potassium Sulfate Solution (Digestion Mixture)*: Dissolve 267 g K_2SO_4 in 1300 mL fresh DDW and 400 mL conc H_2SO_4. While the solution is still hot, add 50 mL mercuric sulfate solution (above) and dilute to 2 L with fresh DDW. In order to keep this reagent in solution, it should be stored in a warm place (~20°C).
4. *Sodium Hydroxide-Sodium Thiosulfate Solution*: Dissolve 500 g NaOH and 25 g $Na_2S_2O_2 \cdot 5H_2O$ in fresh DDW and dilute to 1 L.
5. *Boric Acid Solution*: Dissolve 20 g boric acid, H_3BO_3, in fresh DDW and dilute to 1 L.
6. *Nessler's Reagent*: Dissolve 100 g of mercuric iodide (HgI_2) and 70 g potassium iodide (KI) in a small volume of fresh DDW. Add this mixture slowly, with stirring, to a cooled solution of 160 g of NaOH in 500 mL of fresh DDW. Dilute the mixture to 1 L. The solution is stable for at least one year if stored in a Pyrex bottle out of direct sunlight.

7. *Stock Standard Concentrated Ammonium Chloride Solution*: Dry 2–3 g reagent grade NH_4Cl in a 103°C oven for 1 hr. Cool in a desiccator. Dissolve 1.91 g of NH_4Cl in fresh DDW and dilute to 500 mL.

$$1.00 \text{ mL} = 1 \text{ mg } NH_3\text{-N}$$

8. *Glassware Wash, 6 N HCl*: Add 500 mL conc HCl to 500 mL DDW. Cool. Discard when the solution turns yellow.
9. *Semi-Microkjeldahl Digestion Apparatus and Flasks*: Use 100 mL Kjeldahl flasks to fit in the digestion apparatus. The heating elements should provide the temperature range of 365 to 380°C for effective digestion. The apparatus should have a suction outlet or vent to carry off SO_3 and water fumes. CAUTION: *Always* use this apparatus in a hood.
10. *Distillation Apparatus and Flasks*: Use an all-glass unit with 800 mL macrokjeldahl flasks on heating mantles connected to condensers and adaptors so that distillate can be collected.
11. *Collection Flasks*: 500 mL Erlenmeyer flasks permanently marked at the 350 mL level. Pour 350 mL of water by graduated cylinder in each. Mark and reserve for the TKN analysis only.

C. Standardization

1. Standards are run at a concentration of 1 mg/L. Standards do not require distillation, but an undistilled blank must be run as well as a distilled blank.
2. Standard Curve
   a. Dilute 5.0 mL of the 1 mg/mL stock up to 50 mL; the solution is 100 mg/L.
   b. Dilute 2.0 mL of the 100 mg/L solution up to 200 mL; the solution is 1 mg/L.
3. Pour 50 mL duplicates of the standard. Continue as described in 4.18 D, step 10.
4. With a 1 cm cell, the following gives a guideline for the absorbances to expect for standards and blanks:

Sample	Range of Absorbance at 425 nm
Distilled Blank	0.026–0.042
Undistilled Blank	0.026–0.038
1 mg/L Standard	0.187–0.209

The above ranges assume the use of matched cells.

D. Procedure

Note: Wash all glassware with 6 N HCl and rinse 5–6 times with fresh DDW.

1. Place 50 mL of total (unfiltered) sample into a 100 mL microkjeldahl flask. Add 10 mL of digestion mixture and about 10 boiling chips (Tamer Tabs).
2. Place the flask on the microkjeldahl apparatus (in a hood) and evaporate the mixture at low heat until about one-half of the sample is remaining. Increase heat, and evaporate until SO_3 fumes are given off and the solution turns colorless or pale yellow. Time and continue heating for an additional 30 min. Cool the residue.
3. Transfer the residue quantitatively (using 500 mL of DDW) into an 800 mL Kjeldahl distillation flask. Include the boiling chips also. Cover the flasks with aluminum foil and place on a carrying rack.
4. The distillation apparatus must be presteamed before use, if the system has not been used for 4 hr or more. Prepare about 2 L of a 1 + 1 mixture of fresh DDW and sodium hydroxide-sodium thiosulfate solution. Divide the mixture into two distillation flasks and add boiling chips. Connect to the distillation apparatus with the condenser tips in two collecting flasks. Apply heat and distill until about 100 mL is collected in the flasks. Repeat the above procedure until all units on the distillation apparatus are cleaned.

   Note: If the apparatus has not been used for an extremely long time, it may be necessary to test the distillate produced to see that it is ammonia-free. Proceed to step 10. The absorbance of the sample should be similar to that of a distilled blank.

5. Pour 50 mL of 2% boric acid into each of the marked collection flasks. Position these flasks at the bottom of the distillation apparatus so that the condenser tips (or an extension of them) are below the level of the boric acid solutions.

6. *Note*: Complete this step immediately before placing the distillation flasks on the presteamed apparatus.

   Make the digestate alkaline by careful addition of 10 mL of sodium hydroxide-thiosulfate solution without mixing. Slow addition of the heavy caustic solution down the tilted neck of the digestion flask will cause this heavier solution to underlie the aqueous sulfuric acid solution without loss of free-ammonia nitrogen. Do *not* mix until the digestion flask has been connected to the distillation apparatus.
7. Connect the Kjeldahl flasks to the condensers on the distillation apparatus. Swirl to mix in the heavy caustic solution.
8. Apply heat and distill at the rate of 5–10 mL/min. After 30–60 min, about 300 mL will be distilled over into the collection flask. At this point, drop the collection flask to the lower level on the apparatus. Continue distillation and allow the last 50 mL to drip into the flask.
9. Remove the collection flask when the level reaches the 350 mL mark. Cover the top of the flask with aluminum foil.
10. Measure 50 mL of distillate (or blank, or standard) and add 2 mL of Nessler's reagent. Cover the tops with aluminum foil. The absorbance of the samples must be read after exactly 20 min. This exact timing necessitates the addition of Nessler's to the sample at some interval (1 min) so as to allow spectrophotometric analysis at the correct time.
11. Read the absorbance of the samples in the spectrophotometer at 425 nm using a 1 cm cell.

E. Calculation

Calculate the total Kjeldahl nitrogen in the sample as follows:

$$\text{TKN mgN/L} = (\text{Abs}_{\text{sample}} - \text{Abs}_{\text{dB}})(\text{m}^{-1})(\text{df})$$

where

$$\text{m}^{-1} = \frac{\Delta\text{concentration}}{\Delta\text{absorbance}} = \frac{1 - 0}{\text{Abs}_{\text{std}} - \text{Abs}_{\text{uB}}}$$

UB = undistilled blank
dB = distilled blank

The df is generally 7 because the initial sample volume of 50 mL is diluted to 350 mL during the distillation.

F. Notes

1. Total Kjeldahl nitrogen is defined as the sum of free-ammonia and organic nitrogen compounds converted to ammonium sulfate $(NH_4)_2SO_4$, under the conditions of digestion described.
2. Organic Kjeldahl nitrogen is defined as the difference obtained by subtracting the free-ammonia value from the total Kjeldahl nitrogen value.
3. The Kjeldahl method determines nitrogen in the trinegative state. It fails to account for nitrogen in the form of azide, azine, azo, hydrazone, nitrate, nitrite, nitrile, nitro, nitroso, oxime, and semicarbazone.

4.19 Total Nitrogen Persulfate Method

A. General

1. References: See Solorzano and Sharp (1980, pp. 751–754) and Pitts and Adams (1987, pp. 849–858).
2. Outline of Method:
   Potassium persulfate and heat oxidize nitrogen species to nitrate. Subsequent analysis is for nitrate. Conditions of pH are critical in various steps of the procedure.
   A number approximately equivalent to total Kjeldahl nitrogen can be calculated by subtracting the concentration of nitrate and nitrite in an untreated sample from the concentration of total nitrogen determined in this procedure.

B. Special Reagents and Equipment

1. *Fresh Doubly Dionized Water (DDW)*: Always use fresh DDW from the purification system.
2. *Sodium Hydroxide, 1.5M*: Dissolve 120 g NaOH in 2 L DDW. The solution is stable for months when stored in a tightly closed polypropylene bottle. *Use low-nitrogen, analytical grade NaOH.*
3. *Oxidizing Solution*: Dissolve 6.0 g twice recrystallized potassium persulfate ($K_2S_2O_8$) in 100 mL of 1.5M NaOH, stirring with a Teflon®* coated stirring bar. This will remain stable one week if stored in Pyrex or Teflon.
4. *Hydrochloric Acid, 1.4M*: Dilute 200 mL conc HCl to 1.7 L with DDW. This will be calibrated as outlined in 4.19D.
5. *Buffer Solution*: Dissolve 75 g NH_4Cl in 400 mL DDW. Adjust to pH 8.5 with conc NH_4OH and dilute to 500 mL with DDW. This will remain stable for months when stored in a tightly stoppered glass bottle.

* Registered trademark of E. I. du Pont de Nemours and Company, Inc., Wilmington, Delaware.

6. *Nitrate Stock Standard*: Dry approximately 1 g anhydrous KNO_3 in a 103°C oven for 1 hr. Cool to room temperature in a desiccator. Weigh 0.3609 g and dissolve in DDW. Dilute to 500 mL. The solution is 100 mg NO_3-N/L.
7. *Ammonia Stock Standard*: Dry approximately 2–3 g reagent grade NH_4Cl in a 103°C oven for 1 hr. Cool to room temperature in a desiccator. Weigh 1.91 g and dissolve in DDW. Dilute to 500 mL. The solution is 1000 mg NO_3-N/L.
8. *Reaction Vessels*: Use 125 mL capacity Teflon® bottles (fluorinated ethylene propylene or FEP) or autoclavable plastic bottles (linear polyethylene or LPE; polypropylene or PP). The Teflon® or plastic bottles should be dedicated to this analysis only.

C. Standardization

1. The *Standard Curve* is linear from 0.04 to 10 mg N/L.
2. *Nitrate Standard*: Dilute 1.0 mL stock nitrate solution up to 100 mL with fresh DDW. The solution is 1.0 mg NO_3^--N/L.
3. *Ammonia Standard*: Dilute 1.0 mL stock ammonia solution up to 100 mL with fresh DDW. From this solution, dilute 10.0 mL up to 100 mL with fresh DDW. The final solution is 1.0 mg NH_3-N/L.
4. *Blanks*:
   a. Reagent blank: Add exactly 6.0 mL of oxidizing solution to a 125 mL Teflon® or plastic reaction vessel and autoclave as for samples. After cooling, add 40 mL of DDW and proceed as described below for sample analysis.
   b. Acid calibration blank: Add 40 mL DDW to each of two reaction vessels. Add 6.0 mL oxidizing solution to each and autoclave as described below. Calibrate by following the instructions below.

D. Procedures

1. Rinse plastic or Teflon® reaction vessels with 6 N HCl and fresh DDW 4–5 times. Measure 40 mL of sample into the plastic or Teflon® bottle. Add exactly 6.0 mL oxidizing solution. Autoclave with the cap loose for 30 min at 121°C (15-20 psi). Cool the samples.
2. Calibrate the 1.4M HCl using the acid calibration blanks. Titrate the blanks with the 1.4M HCl to a pH of 2.6–3.2. Dilute the proper amount of 1.4M HCl so that 6.0 mL gives a pH of 2.6–3.2 (the solution will be $\approx$ 1.0M).

3. Add exactly 6.0 mL of the calibrated HCl solutions to the samples to dissolve any precipitate and to lower the pH. Quantitatively transfer the solution to 125 mL Erlenmeyer flasks. Add 3.0 mL of buffer solution to the Erlenmeyer flask. (Do not add the buffer to the Teflon® bottles).
4. Analyze the sample for nitrate using a cadmium-reduction technique (see 4.16, 4.17).
5. To obtain a number comparable to TKN, also analyze untreated samples for nitrate and nitrite.

E. Calculation

$$\text{Total Nitrogen, mg N/L} = (A - B)(1.32)$$

$$\text{TKN, mg N/L} = (A - B)(1.32) - C$$

where

A = mg NO_3^--N/L in digested sample
B = mg NO_3^--N/L in reagent blank
C = mg (NO_3^--N + NO_2^--N)/L in untreated sample
1.32 = df

F. Notes

The standards are set up as a check on the procedure.

4.20 Dissolved Oxygen (DO) Winkler with Azide Modification Method

A. General

1. References: See *Standard Methods* (1989 pp. 4–149 through 4–156) and EPA (1983, Method 360.2).
2. Outline of Method:
   A divalent manganese solution followed by a strong alkali is added to the sample. Any dissolved oxygen (DO) rapidly oxidizes an equivalent amount of divalent manganese to basic hydroxides of higher valence states. When the solution is acidified in the presence of iodide, the oxidized manganese again reverts to the divalent state and iodine, equivalent to the original dissolved oxygen content of the water, is liberated. The amount of iodine is then determined by titration with standardized thiosulfate solution.

B. Special Reagents

1. *Manganese Sulfate Solution*: Dissolve 364 g $MnSO_4 \cdot H_4O$ in DDW and dilute to 1 L; filtration of the reagent may be necessary if dissolution is not complete.
2. *Alkali-Iodide-Azide Reagent*: Dissolve 500 g of solid NaOH and 135 g NaI (sodium iodide) in DDW and dilute to 1 L. Add to this solution 10 g NaN_3 (sodium azide) dissolved in 40 mL DDW.
3. *Concentrated Sulfuric Acid*: About 36 N H_2SO_4. Hence, 1 mL is equivalent to about 3 mL of the alkali-iodide-azide reagent.
4. *Starch Solution*: Mix 10 g of laboratory grade soluble starch and a few mL of DDW in a beaker or mortar. Pour this emulsion into 400 mL of boiling DDW. Dilute to about 500 mL. Allow it to boil a few minutes and let it settle overnight. Use the clear supernatant. Store in a plastic squeeze bottle in the refrigerator. *This will remain stable one month*; discard the solution when the endpoint color is no longer pure blue, but takes on a green or brown tint. This solution may be preserved with 1.0 g salicylic acid.
5. *Sodium Thiosulfate Stock Solution, 0.10 N*: Dissolve 24.82 g $Na_2S_2O_3 \cdot 5H_2O$ in boiled and cooled DDW and dilute to 1 L. Preserve the solution by adding 5 mL chloroform or 1 g NaOH per liter.

6. *Standard $Na_2S_2O_3$ Titrant, 0.0250 N*: Dilute 250 mL stock standard to 1 L; exactly 1.00 mL 0.0250 N is equivalent to 200 μg/L D.O.
7. *Standard Potassium Dichromate, 0.025 N*: Dry approximately 2–3 g $K_2Cr_2O_7$ at 103°C for 2 hr and then dissolve 1.226 g of the dried $K_2Cr_2O_7$ in DDW and dilute to 1 L.
8. *Sulfuric Acid, 10% V/V*: Carefully add 50 mL conc H_2SO_4 to 300 mL DDW. Cool and dilute to 500 mL.
9. *Potassium Iodide*: Use (KI) crystals.

C. Standardization

Thiosulfate Standardization: Dissolve approximately 2 g KI (free of iodate) in 150 mL DDW in a 500 mL Erlenmeyer flask. Add 10 mL of 10% V/V H_2SO_4 followed by exactly 20 mL of standard 0.025 N $K_2Cr_2O_7$. Place in the dark for 5 min, dilute to approximately 400 mL, and titrate with 0.025 N thiosulfate solution.

$$\text{N of } Na_2S_2O_3 = \frac{(0.025)\ (20)}{\text{mL } Na_2S_2O_3 \text{ used}}$$

D. Procedure

1. Rinse a 300 mL BOD bottle with sample. Pour the sample into the BOD bottle using a reversing sampler with a length of rubber tubing that extends from the top to the bottom of the bottle. The end of the rubber tube must remain beneath the surface of the water as the bottle is filled. Water is allowed to overflow from the top of the bottle (at least 1/3 of the volume of the bottle should be allowed to overflow). The bottle is then stoppered when all the air bubbles, if any, have been allowed to rise out of the BOD bottle. The temperature of the sample should be recorded.
2. Remove the glass stopper and add 2 mL of $MnSO_4$ reagent followed by 2 mL of the alkali-iodide-azide reagent; introduce both these reagents beneath the surface of the sample. Replace the stopper, being careful not to trap air inside. Mix by inverting the bottle at least 15 times. Allow floc to settle and invert again. Allow floc to settle again and remove the stopper. Immediately add 2 mL conc H_2SO_4 by allowing the acid to run down the neck of the bottle, restopper, and mix until the precipitate dissolves, leaving a clear yellow-orange iodine solution. Dissolution should be complete. Samples stored at this point should be protected from strong sunlight and titrated as soon as possible (within 4 to 8 hr).

3. Measure 203 mL of sample (this corresponds to 200 mL of the original sample) into a 250 mL Erlenmeyer flask.
4. Rinse the burette with fresh 0.025 N $Na_2S_2O_3$ and then titrate to a faint yellow color (use a white background). Add 1–2 mL of the starch solution and continue the titration until the solution changes from blue to clear.

   Note: This titration must not be delayed and the thiosulfate should be added fairly rapidly. Solutions should remain colorless for at least 20 sec at the endpoint.

   Note: Use of the starch solution facilitates clear endpoint detection by forming a blue complex with any iodine remaining in the solution.

5. 1.0 mL of 0.025 N $Na_2S_2O_3$ is equivalent to 200 μg DO/L; therefore, if a 203 mL sample (200 mL of original sample) is titrated, 1.0 mL 0.025 N $Na_2S_2O_3$ equals 1 mg O_2/L as DO.

E. Calculation

$$\text{DO} = \text{mL of Na}_2\text{S}_2\text{O}_3 \text{ used} \times \frac{\text{N}}{0.025}$$

F. Note

The $Na_2S_2O_3$ should be standardized fairly frequently for normality changes so that appropriate correction of measured DO can be made.

4.21 Reactive (Ortho) Phosphate
Ascorbic Acid or Murphy-Riley Method

A. General

1. Reference: See *Standard Methods* (1989, pp. 4–166 through 4–170; 4–177 through 4–178).
2. Outline of Method:
   Reactive phosphate in the sample is complexed in the presence of molybdic acid, ascorbic acid, and antimony to a blue-colored molybdate complex. The absorbance intensity of the blue color is measured at 880 nm.

B. Special Reagents

1. *Armstrong Reagent*: Add 122 mL of conc H_2SO_4 to 800 mL of DDW. While the solution is still hot, add 10.5 g of ammonium molybdate and 0.3 g antimony potassium tartrate. Heat to dissolve, cool, and dilute to exactly 1 L with DW. This will remain stable indefinitely.
2. *Ascorbic Acid Solution*: Dissolve 3 g of ascorbic acid in 100 mL of DDW. Note: This will remain stable only *one week* (store in the refrigerator). Be certain to date the reagent bottle each time the solution is prepared.
3. *Hydrochloric Acid, 6 N*: Carefully add 500 mL conc HCl to 500 mL DDW. Cool. Reuse and discard when the solution turns yellow.
4. *Stock Potassium Phosphate*: Dry 3–4 g of reagent grade anhydrous K_2HPO_4 at 103°C for 1 hr. Dissolve exactly 2.81 g of the dry salt in DDW and dilute to exactly 500 mL with DDW. Store in a plastic bottle. (This will remain stable for six months.)

$$1.00 \text{ mL} = 1 \text{ mg } PO_4^{-3} - P$$

C. Standardization

1. Standard Curve

$$10 \text{ to } 700\ \mu g\ P/L$$

2. Procedure for daily standardization
   a. Dilute 1.0 mL of the 1 mg/mL stock PO_4^{-3}-P solution up to 100 mL; solution is 10 mg/L.
   b. Prepare one of the following standards. Choose a standard that is closest to your sample concentrations.

Final Concentration	mL of the 10 mg/L Solution
100 μg/L	2.0 mL diluted up to 200 mL
200 μg/L	4.0 mL diluted up to 200 mL

3. Pour 50 mL duplicates of the standard. Also, pour two 50 mL DDW blanks. Treat standards and blanks exactly like the samples in the procedure below.
4. With a 5 cm cell at 880 nm, the absorbance of the 100 μg/L standard minus the blank should be in the range of 0.298–0.320. If the standard's absorbance does not fall within this range, it is time to start looking for problems with the analysis.

D. Procedure

1. Rinse all glassware with 6 N HCl and then 4–5 times with DDW.
2. Measure a 50 mL sample into a 125 mL Erlenmeyer flask. Use filtered sample only.
3. Add 5 mL of Armstrong reagent and 1 mL of ascorbic acid; swirl to mix.
4. Allow a 20 min reaction period (it will remain stable up to 2 hr), and then read the absorbance against DDW at 880 nm using a 1 or 5 cm cell. Record the cell used.
5. If necessary, measure absorbance of the samples by adding Armstrong's reagent (but *not* the ascorbic acid) to obtain a turbidity correction factor.
6. Subtract turbidity correction (if any) from sample absorbance. Plug this corrected absorbance into the slope-intercept formula for the standard curve to obtain μg PO_4^{-3}-P/L.

E. Calculation

Calculate the reactive (ortho) phosphate in the sample as follows:

$$\mu g\ PO_4^{-3} - P/L = (Abs_{sample} - Abs_{blank})(m^{-1})(df)$$

where

$$m^{-1} = \frac{\Delta concentration}{\Delta absorbance} = \frac{100 - 0}{Abs_{std} - Abs_{blank}} \text{ or } \frac{200 - 0}{Abs_{std} - Abs_{blank}}$$

df = dilution factor

If a 10% sample is used, df = 10; if a 5% sample is used, df = 20.

4.22 Total Phosphorus

A. General

1. Reference: See *Standard Methods* (1989, pp. 4–166 through 4–172; 4–177 through 4–178).
2. Outline of Method:
   The total phosphorus content of the sample includes all the orthophosphate, polyphosphates, and most typical organic phosphate compounds that are hydrolyzed to orthophosphate through mild acid hydrolysis. If knowledge about more complete digestion is required, compare this method with other digestion techniques (see *Standard Methods*, pp. 4–171 through 4–172). Polyphosphates do not respond appreciably to the orthophosphate tests alone, but are hydrolyzed to orthophosphate by boiling with acid.
   After digestion, total phosphorus is then measured quantitatively by the method selected for orthophosphate determinations (ascorbic acid).

B. Special Reagents

1. *Strong Acid for Hydrolysis*: Slowly add 300 mL of conc H_2SO_4 to approximately 600 mL of DDW. When cooled, dilute to 1 L.
2. *Phenolphthalein Indicator*: Dissolve 2.5 g of phenolphthalein powder into 250 mL of ethyl alcohol and dilute to 500 mL with DDW.
3. *Sodium Hydroxide, 1 N*: Dissolve 40 g solid NaOH in DDW and dilute to 1 L.
4. *Ammonium Persulfate, $(NH_4)_2S_2O_8$*: Chemical reagent and small powder scooper ($\sim$0.4g capacity).
5. *Armstrong Reagent*: See 4.21.
6. *Ascorbic Acid*: See 4.21.
7. *Hydrochloric Acid, 6 N*: See 4.21.
8. *Stock Potassium Phosphate*: See 4.21.

C. Standardization

1. See 4.21. Set up the 200 μg/L standard.
2. Pour 50 mL duplicates of the standard and pour two 50 mL DDW blanks.

3. Treat the standards and blanks exactly like the samples for the digestion and analysis procedure below.
4. The following absorbance ranges can be expected for the standards. If the absorbance of the standard does not fall within the range, it is time to start looking for problems with the analysis.

Standard μg/L	Range of Absorbances of (Std-Blank) Using 5 cm Cell at 880 nm
100	0.151–0.174
200	0.271–0.318

D. Procedure

1. Rinse glassware with 6 N HCl and rinse 4–5 times with DDW.
2. Add 1 mL of the strong-acid solution to a 50 mL sample in a 125 mL Erlenmeyer flask.
3. Add 0.4 g ammonium persulfate. Cover the flask with an inverted beaker or aluminum foil.

   Caution: Ammonium persulfate is a strong oxidant. Handle it with care.

4. Heat in the autoclave for 30 min at 121°C (15–20 psi).
5. If suspended material is present in the sample at this time, filter the sample before neutralization. Use GF/C glass fiber filters.
6. Add 4–5 drops of phenolphthalein and adjust to pink with 1 N NaOH. Bring back to colorless with one drop of the strong-acid solution. Cool.
7. Dilute to 100 mL with DDW using a volumetric flask.
8. Measure out 50 mL of sample and determine orthophosphate using the ascorbic acid technique. The remaining sample may be saved in case mistakes are made or a dilution is necessary.

E. Calculation

See 4.21.

4.23 Total Phosphorus in Sediments
Persulfate Digestion Method

A. General

1. References: See *Standard Methods* (1989, p. 4–172) and Schmalz (1971, pp. 52–54).
2. Outline of Method:
   A sediment sample is oxidized in an acidic solution in order to convert all phosphorus forms to orthophosphate. The sample is then filtered and the phosphorus measured quantitatively using the ascorbic acid method.

B. Special Reagents

1. See 4.21 and 4.22.

C. Standardization

1. See 4.21. Set up the 200 μg P/L standard.

D. Procedure

1. Weigh out 0.15 g of oven-dried (8–10 hr at 103°C) sediment and place in a 125 mL Erlenmeyer flask.
2. Add 50 mL DDW, 1 mL strong-acid solution, and 0.4 g $(NH_4)_2S_2O_8$. Also run two 50 mL DDW blanks and two 200 μg P/L standards. Cover flasks with inverted beakers or aluminum foil.
3. Digest the mixture in an autoclave for 30 min at 121°C (15–20 psi); cool.
4. Filter the sample through a GF/C filter. Rinse the flask well with small portions of DDW, pouring the washings through the filter also.

   Note: Be careful not to exceed a final volume of 100 mL. The filtrate should be clear.

5. Transfer the filtrate quantitatively to an Erlenmeyer flask. Add 4–5 drops of phenolphthalein indicator and adjust filtrate to a pink color

with 1 N NaOH. Bring back to colorless with one drop of strong acid solution. Transfer quantitatively to a 100 mL volumetric flask and dilute to volume with DDW.

Note: Be careful *not* to add excess base here. This may cause phosphorus to precipitate out of solution.

6. Measure 50 mL of sample (diluted to 100 mL) and determine orthophosphate using the ascorbic acid technique (see 4.21). The remaining sample may be saved in case mistakes are made or dilutions are necessary.

E. Calculation

1. *Procedure in brief*

0.15 g sediment → 50 mL DDW + reagents
↓
Autoclave
↓
Filter
↓
Neutralize and Dilute to 100 mL
Save Remaining Solution ← ↓
Analyze 50 mL: $\text{Absorbance}_{\text{sample}}$

2. *Derivation of calculation*

a. $(\text{Abs}_{\text{sample}} - \text{Abs}_{\text{blank}}^{*})(\text{m}^{-1})$ = concentration for 50 mL subsample in μg P/L

where

$$\text{m}^{-1} = \frac{\Delta\text{concentration}}{\Delta\text{absorbance}} = \frac{200 - 0}{\text{Abs}_{\text{std}} - \text{Abs}_{\text{blank}}^{*}}$$

b. $(\text{Abs}_{\text{sample}} - \text{Abs}_{\text{blank}}^{*})(\text{m}^{-1})(1/20)$ = concentration for 50 mL subsample in μgP/50 mL

where

* This is the blank run through the procedure.

$$\frac{1}{20} = \frac{50 \text{ mL}}{1000 \text{ mL}}$$

c. $(Abs_{sample} - Abs_{blank}{}^{*})(m^{-})(1/20)(1/0.075g) = \text{concentration in} \frac{\mu gP}{\text{g sediment}}$

where

$$\frac{1}{0.075g} = \text{sediment for 50 mL subsample}$$

3. *Simplified form of calculation*

$$\frac{\mu gP}{\text{g sediment}} = \frac{(Abs_{sample} - Abs_{blank}{}^{*})(m^{-1})}{1.5}$$

* This is the blank run through the procedure.

4.24 Silica
Molybdosilicate Method

A. General

1. References: See *Standard Methods* (1989, pp. 4–181 through 4–187), and EPA (1983, Method 370.1).
2. Outline of Method:
   Silica (along with phosphates) will react with ammonium molybdate at ~pH 1.2 to form heteropoly acids. Oxalic acid is then added to destroy any molybdophosphoric acid present. The molybdosilicic acid (yellow in color) solution is measured spectrophotometrically at 410 nm. Store samples in plastic; avoid glass as much as possible in the procedure. Interferences are iron, color, turbidity, S^{-2}, and phosphate.

B. Special Reagents

Store all reagents in polyethylene containers.

1. *Hydrochloric Acid, 1 + 1*: Mix equal parts of HCl and DDW.
2. *Ammonium Molybdate Reagent*: Dissolve 10 g $(NH_4)_6Mo_7O_24 \cdot 4H_2O$ in DDW with stirring and gentle warming, and dilute to 100 mL. Filter if necessary. Adjust pH to 7–8 with 1 N NaOH to stabilize.
3. *Oxalic Acid*: Dissolve 10 g $H_2C_2O_4 \cdot 2H_2O$ in DDW and dilute to 100 mL.
4. *Silica Dilut-it Stock, 1 mg SiO_2/mL*: Prepare a 1 mg SiO_2/mL stock solution using a silica Dilut-it concentrated standard. Follow the manufacturer's instructions.

C. Standardization

1. Concentration range: 1 to 40 mg SiO_2/L.
2. Daily procedure: Dilute 1.0 mL of the stock standard up to 100 mL with DDW.

10 mg SiO_2/L solution

Set up two 10 mg/L standards and two DDW blanks. Treat standards and blanks exactly like samples in the procedure below.

3. The absorbance values to be expected for the standards minus the blank will vary somewhat as batches of reagents are prepared. Keep good records of data to determine acceptable absorbance ranges for the standards and blanks.

D. Procedure

1. Pour samples, standards, and blanks in plastic containers with covers. Use a 50 mL sample or an aliquot diluted to 50 mL. In rapid succession add 1.0 mL of 1 + 1 HCl and 2.0 mL of the ammonium molybdate solution.
2. Mix by inverting at least six times.
3. Allow the solution to stand 5–10 min.
4. Add 1.5 mL of the oxalic acid solution and mix thoroughly.
5. After 2 min (but before 15 min have elapsed after the addition of oxalic acid), read the color on a spectrophotometer at 410 nm.

E. Calculation

$$\text{mg } SiO_2/L = (Abs_{sample} - Abs_{blank})(m^{-1})(df)$$

where

$$m^{-1} = \frac{\Delta concentration}{\Delta absorbance} = \frac{10 - 0}{Abs_{std} - Abs_{blank}}$$

$$df = \text{dilution factor}$$

If a 10% sample is used, df = 10; if a 5% sample is used, df = 20.

4.25 Sulfate
Turbidimetric Method

A. General

1. References: See *Standard Methods* (1989, pp. 4-204 through 4-208) and EPA (1983, Method 375.4).
2. Outline of Method:
   Sulfate ion forms a uniform precipitate in a hydrochloric acid medium with barium chloride. The absorbance of the resultant barium sulfate suspension is measured spectrophotometrically. Analyzed at room temperature, under strongly acidic conditions, sulfate is the only ion that will form insoluble compounds in normal waters. The accuracy of the technique decreases for sulfate concentrations above 30 mg/L, so dilution with DDW may be necessary to obtain the appropriate concentration range.

B. Reagents and Special Equipment

1. *Conditioning Reagent*: Mix 50 mL glycerol with a solution of 30 mL conc HCl, 300 mL DDW, 100 mL 95% ethyl or isopropyl alcohol, and 75 g sodium chloride.
2. *Barium Chloride*: Use crystals, 20–30 mesh.
3. *Stock Sulfate Solution*: Dissolve 1.479 g anhydrous sodium sulfate Na_2SO_4 in DDW and dilute to 1 L.

$$1.00 \text{ mL} = 1000 \ \mu\text{g } SO_4^{=}$$

4. *Magnetic Stirrer, Variable Speed*:

   Note: Use magnetic stirring bars of identical shape and size.

5. *Stopwatch*: An accurate timer may also be used.

C. Standardization

1. Concentration range:

$$1 \text{ mg } SO_4^{=}/L \text{ to } 30 \text{ mg } SO_4^{=}/L$$

2. Daily procedure:

Make 100 mL standards in the following concentrations.

Final Concentration, mg $SO_4^=/L$	mL Stock Sulfate Solution
10	1.0 mL diluted up to 100 mL
20	2.0 mL diluted up to 100 mL
30	3.0 mL diluted up to 100 mL

Also, set up one 100 mL blank (DDW). Treat standards and the blank exactly like the samples in the procedure below.

D. Procedure

1. Measure a 100 mL sample (or an aliquot diluted to 100 mL) into a 250 mL Erlenmeyer flask.
2. Add exactly 5.00 mL of conditioning reagent.
3. Mix, using the magnetic stirrer and stirring bar.

   Note: Always set magnetic stirrer at the same speed and use stirring bars of identical shape and size.

4. While the solution is stirring, add a small scoop (0.2–0.3 g) of $BaCl_2$ crystals and begin timing immediately.
5. Stir exactly 1 min at a constant speed.
6. At the end of the stirring period, place the sample in a 5 cm cuvette and measure the absorbance at 420 nm after exactly 4 min.
7. Run a 20 mg $SO_4^=/L$ standard every 10–12 unknowns to insure that conditions are stable.
8. If any turbidity is present, use the absorbance of sample-plus-conditioning reagent as the blanks for that particular sample.

E. Calculation

Calculate the concentration of $SO_4^=/L$ in a sample using a linear regression of the calibration standards 1–30 mg/L range; or plot the absorbance of the calibration standards against the calibration concentrations and compute the sample concentration directly from the linear standard curve.

4.26 Sulfate
Gravimetric Method with Drying of Residue

A. General

1. References: See *Standard Methods* (1989, pp. 4–204 through 4–206) and EPA (1983, Method 375.3).
2. Outline of Method:
   Sulfate is precipitated as barium sulfate in hydrochloric acid medium by the addition of barium chloride. The $BaSO_4$ precipitate is washed until it is free of chloride, dried, and weighed.

   This method is subject to many positive and negative interferences. Factors leading to high results include suspended matter, silica, $BaCl_2$ precipitant, nitrate, sulfite, and occluded water particles. A factor leading to low results is the presence of alkali metal sulfates.

   Use this method only when sulfate concentrations are very high (greater than 300 mg $SO_4^{=}$/L) and interferences are expected to be of minor importance. Consult the above references for methods to remove interferences.

B. Special Reagents and Equipment

1. *Methyl Red Indicator*: Dissolve 100 mg of methyl red sodium salt in DDW and dilute to 100 mL.
2. *Hydrochloric Acid, 6 N*: In a hood, carefully add 500 mL of conc HCl to 500 mL DDW with stirring. Cool.
3. *Barium Chloride Precipitating Solution*: Dissolve 100 g $BaCl_2 \cdot 2H_2O$ in DDW and dilute to 1 L. Filter through a GF/C or membrane filter. One mL of this solution will precipitate approximately 40 mg $SO_4^{=}$.
4. *Silver Nitrate-Nitric Acid Reagent*: Dissolve 8.5 g $AgNO_3$ in DDW. Add 0.5 mL conc HNO_3 with DDW. Dilute to 500 mL.
5. *Sodium Sulfate Standard*: Dry a small quantity of anhydrous $NaSO_4$ in a 103°C oven for 1 hr. Cool in a desiccator.
6. *Autoclave or Steam Bath*: The autoclave must be capable of maintaining temperature in the range of 80 to 90°C and have multi-therm capabilities.
7. *Filter*: The filter should have a fine glass frit (maximum pore size 5 μm) and the capability of being fitted to a vacuum filtering flask.

8. *Filter Paper*: GF/C filter paper (or the equivalent) is needed to fit filtering apparatus well.
9. *Drying Oven*: The oven must be maintained at 105°C or higher.
10. *Analytical Balance*: The balance must be capable of weighing to the nearest 0.1 mg.
11. *Watch Glasses*: These must be of sufficient size to loosely cover a 500 mL beaker.

C. Standardization

A standard is not required in this procedure. However, running a standard through the procedure is a good check on the reliability of digestion, filtering, and drying conditions.

Weigh accurately two 60 mg quantities of dried and cooled Na_2SO_4 into two clean 500 mL beakers. Record exact quantities weighed to the nearest 0.1 mg. This weight of Na_2SO_4 will produce about 100 mg of $BaSO_4$ precipitate. Add about 250 mL DDW to dissolve (the volume is not too important). Treat as samples in the procedure below. Measure two 250 mL blanks into two 500 mL beakers. Treat as samples in the procedure below.

D. Procedure

1. Consult Bibliography for the removal of interferences if necessary.
2. Accurately measure enough sample to contain about 50 mg sulfate ion into a 500 mL beaker. Do not allow sample volume to exceed 150 mL.

   Note: Samples with sulfate concentrations of less than about 300 mg $SO_4^=$/L are not applicable to this procedure. Refer to 4.25.

3. Add enough DDW to bring the total volume up to about 250 mL.
4. Adjust the pH with 6 N HCl to 4.5–5.0 using a pH meter or the orange color of methyl red indicator.
5. Add an additional 2.0 mL of 6 N HCl.

   Note: The pH is lowered here to prevent the precipitation of barium carbonate and barium phosphate. However, at this pH, $BaSO_4$ has a small but significant solubility. Therefore it is important to limit the concentration of ionized $BaSO_4$ and keep the acid concentration consistent.

6. Heat the solution just to boiling and add 5 mL of warm $BaCl_2$ precipitating solution slowly while stirring gently with a glass rod. Precipitation should be complete; the last 2 mL of $BaCl_2$ should not produce any noticeable precipitate formation.
7. Rinse the glass rod with a small amount of DDW into the beaker.
8. Cover the beakers with loose-fitting watch glasses.
9. Digest the solutions in an autoclave or steam bath at 80 to 90°C for at least 2 hr but preferably overnight. Cool to room temperature.
10. Place the GF/C filters in fritted glass filter setups. Make certain that the filters fit well. Attach the filter setup to a vacuum source and rinse well with DDW.
11. Dry the filter setups to constant weight in an oven maintained at 105°C or higher.
12. Cool in a desiccator and weigh.
13. Filter the samples at room temperature through the preweighed filter setups. Use DDW to rinse all of the precipitate out of the beaker.
14. Rinse the precipitate in the filter setup with several small portions of warm DDW. Test the washings in the vacuum filter flask for the presence of chloride by adding a few drops of $AgNO_3$-HNO_3 reagent. If chloride is present, the washings will turn milky or cloudy. Discard these washings and continue rinsing with warm DDW until the washings are free of chloride.
15. Dry the filter setup and precipitate to constant weight, using exactly the same procedure used in preparing the filter.
16. Cool the filter setups in a desiccator, reweigh, and record the results.

E. Calculation

1. For each sample, standard, or blank, calculate the weight of the $BaSO_4$ precipitate found.

$$\text{mg } BaSO_4 = (\text{weight of filter + precipitate, mg}) - (\text{weight of filter, mg})$$

2. Subtract the average mg $BaSO_4$ found in the blanks from the mg $BaSO_4$ found in each of the standards and samples. Use these blank-corrected data in all the calculations below.
3. Calculate the percent recovery of the standards as follows:

$$\text{percent recovery} = \frac{A}{B \times 1.64} \times 100$$

where

A = mg $BaSO_4$ found in the standard
B = mg Na_2SO_4 weighed

If the percent recovery is $\leq 90\%$ or $\geq 100\%$, take corrective action to determine the cause of the low results. (See notes below.)

4. If the percent recovery of the standard is 90 to 100%, calculate the results as follows:

$$\text{mg } SO_4^{=}/L = \frac{C \times 411.6}{D}$$

where

C = mg $BaSO_4$ precipitate formed
D = sample volume in mL

F. Notes

If results for the standard are high, occluded water in the precipitate may be the cause. Occlusion means that water or the mother liquor has been physically trapped in $BaSO_4$ precipitate itself. If occlusion is a continuing problem, consult the above references and use the ignition method, igniting the residue at 800°C instead of drying it in an oven.

4.27 Sulfite
Titrimetric Method

A. General

1. References: See *Standard Methods* (1989, pp. 4–199 through 4–201) and EPA (1983, Method 377.1).
2. Outline of Method:
   An acidified sample is titrated with standard potassium iodide-iodate titrant. The titrant liberates free iodine that reacts with the sulfite in the sample. When the sulfite has been consumed, the excess iodine forms a blue color with a starch indicator.
   Consult the Bibliography for a list of interfering substances and procedures to remove them.
3. Test for sulfite as soon as possible after the collection of the sample.

B. Special Reagents

1. *Sulfuric Acid, 50%*: In a hood, carefully add 500 mL conc H_2SO_4 to 500 mL DDW with stirring. Cool.
2. *Standard Potassium Iodide-Iodate Titrant, 0.0125 N*: Dry a small amount of primary standard grade anhydrous KIO_3 at 120°C for 4 hr. Cool in a desiccator. Dissolve 445.8 mg KIO_3, 4.35 g KI, and 310 mg sodium bicarbonate ($NaHCO_3$) in DDW and dilute to 1 L. 1.00 mL = 500 μg $SO_3^{=}$.

   Note: The titrant should be clear, not yellow in color. The yellow color is caused by the presence of free iodine and may be due to the abnormally high pH of the DDW (> 7 or 8) or air oxidation of the KI. Air oxidation can occur when the KI reagent is old or has not been stored in a dark bottle.

3. *EDTA Solution*: Dissolve 2.5 g of disodium EDTA in 100 mL DDW. This solution is used as a fixative at the time of sample collection to minimize the effects of oxidizable materials (see above references).
4. *Starch Indicator*: See 4.20.
5. *Sulfamic Acid*: Use crystalline NH_2SO_3H.

C. Standardization

It is not necessary to carry a standard through the procedure. However, two 50 or 100 mL blanks should be carried through the procedure below.

D. Procedure

1. Samples must be freshly collected and below 50°C. Minimize contact with the air at all times. Do not filter. Add 1 mL of EDTA solution for every 100 mL of sample. Analyze immediately.
2. Add 1 mL of 50% H_2SO_4 to a 250 mL titration flask.
3. Add 0.1 g sulfamic acid crystals to eliminate nitrite interference.
4. Measure 50 to 100 mL of sample into the flask. Again, avoid contact with the air; keep pipet tip below the surface of the liquid.
5. Add 1 mL of starch indicator.
6. Titrate immediately with KI-KIO_3 standard titrant while stirring.
7. The endpoint is reached when a faint blue color develops and persists.
8. Record the volume of titrant used.

E. Calculation

Calculate the sulfite in the sample as follows:

$$\text{mg } SO_3^{=}/L = \frac{(A - B) \times N \times 40{,}000}{\text{mL sample}}$$

where

A = mL of titrant required for sample
B = mL of titrant required for blank
N = normality of KI-KIO_3 titrant
Note: This normality is usually 0.0125

Methods for the Determination of Organics

5. Methods for the Determination of Organics

5.1 Biochemical Oxygen Demand (BOD) Method

A. General

1. References: See *Standard Methods* (1989, pp. 5–2 through 5–10), Sawyer and McCarty (1978, pp. 416–432), Young et al. (1981, pp. 1253–1259), and EPA (1983, Method 405.1).
2. Outline of Method:
   The BOD test determines the relative oxygen necessary for biological oxidation of wastewaters, effluents, and polluted waters. It is the only test available to determine the amount of oxygen required by bacteria while stabilizing decomposable organic matter. Complete stabilization requires too long an incubation period for practical purposes; therefore, the 5-day period has been accepted as a standard. Samples are incubated in the dark at 20 ± 1°C. Dissolved oxygen levels are measured initially and at the end of the 5-day period using the Winkler with Azide Modification method. The 5-day incubated sample must deplete more than 2 mg/L dissolved oxygen and have more than 0.5 mg/L dissolved oxygen left.
3. Set up BOD tests as soon as possible after the sample is collected. *Standard Methods* (1989, p. 1–37) suggests a maximum holding time of 6 hr where possible.

B. Special Reagents and Equipment

1. *BOD incubator*: Set at 20 ± 1°C. Record the temperature daily.
2. For dilution water
   a. *Phosphate Buffer Solution*: Dissolve 8.5 g KH_2PO_4, 21.75 g K_2HPO_4, 44.6 g $Na_2HPO_4 \cdot 12\ H_2O$, (33.4 g $Na_2HPO_4 \cdot 7H_2O$) and 1.7 g NH_4Cl in about 500 mL of DDW and dilute to 1 L. Check the pH monthly; it should be 7.4 ± 0.1. Discard the buffer if it is not within this range.
   b. *Magnesium Sulfate Solution*: Dissolve 22.5 g $MgSO_4 \cdot 7H_2O$ in DDW and dilute to 1 L.
   c. *Calcium Chloride Solution*: Dissolve 27.5 g anhydrous $CaCl_2$ (36.4 g $CaCl_2 \cdot 2H_2O$) in DDW and dilute to 1 L.
   d. *Ferric Chloride Solution*: Dissolve 0.25 g $FeCl_3 \cdot 6H_2O$ in DDW and dilute to 1 L.

3. For dechlorination
   a. *Sodium Sulfite Solution, 0.025 N*: Dissolve 787.5 mg anhydrous Na_2SO_3 in 500 mL DDW. Prepare this solution fresh daily.
   b. *Sulfuric Acid Solution, 1→50 V/V*: Add 4.0 mL conc H_2SO_4 to 150 mL DDW. Cool and dilute to 200 mL.
   c. *Potassium Iodide Solution*: Dissolve 10 g KI in 100 mL DDW. Prepare this fresh monthly. Store in the dark. Discard if the solution is yellow.
   d. *Starch*: See 4.20.
4. For pH adjustment
   a. *Sulfuric Acid, 1 N*: Carefully add 28 mL conc H_2SO_4 to 500 mL DDW. Cool and dilute to 1 L.
   b. *Sodium Hydroxide, 1 N*: With stirring, dissolve 40 g NaOH in DDW. Cool and dilute to 1 L.
5. Dissolved oxygen: See 4.20 for method and reagents.
6. Seed: This is a biological population capable of oxidizing organic matter. One type of seed material is settled raw domestic sewage that has been stored at 20°C for 1 to 36 hr. After this initial period at 20°C, store seed in the refrigerator. Preferably, use seed collected from a source most likely to contain organisms that are acclimated to the particular characteristics of the wastewater being analyzed. *Standard Methods* (1989, p. 5-6) suggests seed in the effluent from a biological treatment system processing the waste. Alternately, obtain seed from the receiving waters below the point of discharge.

C. Standardization

See 4.20.

D. Procedure

1. *Undiluted Samples*: These are made up of high quality water whose 5-day BOD is less than 7 mg DO/L. The initial DO should be near saturation levels; if not, aerate the sample to saturation. Measure initial DO immediately. Incubate at 20 ± 1°C for 5 days, and measure the final DO.

$$\text{mg BOD}_5/\text{L} = D_1 - D_5$$

where

D_1 = initial dissolved oxygen
D_5 = 5-day dissolved oxygen

2. *Diluted Samples*: Because of the limited solubility of oxygen in water, samples with suspected high BODs must be diluted.

   Always use high quality laboratory DDW for dilution water. Aerate the water with a supply of clean compressed air. Dilution water should be at 20 ± 1°C. Add 1 mL each of phosphate buffer, magnesium, calcium, and ferric solutions for each liter of dilution water needed. If dilution water is stored, add phosphate buffer just prior to use.

   As a test of the quality of the dilution water, incubate a bottle full of dilution water for 5 days at 20°C. Determine initial (D_1) and 5-day (D_5) DO. The DO uptake in 5 days should not be more than 0.2 mg/L and preferably not more than 0.1 mg/L. If the DO consumption was greater than 0.2 mg/L, the results for the samples using the same dilution water should be rejected. See *Standard Methods* (1989, pp. 5-4 through 5-6), and Young et al. (1981, pp. 1253–1259), for suggestions for the improvement of dilution water quality.

   Prepare three sets of BOD bottles, including dilution water blanks; one set is for the initial determination and the other two are for the 5-day oxygen determination.

   Make several dilutions of prepared sample so as to obtain sufficient oxygen depletions.

0.1–1.0%	for strong trade wastes
1–5%	for raw and settled sewage
5–25%	for oxidized effluents
25–100%	for high concentrations of organic wastes

   See also Table 5.1.

   Prepare dilutions either in 1 L graduated cylinders (and then transfer to BOD bottles), or prepare directly in the BOD bottles. When using graduate cylinders, syphon dilution water carefully, filling the cylinder half full. Add the proper amount of mixed sample to the cylinder, and fill to the 1 L line with syphoned dilution water. Mix well with a

Table 5.1 BOD Measurable with Various Dilutions of Samples

Using Percent Mixture		By Direct Pipetting Into 300 mL - Bottles	
Percent Mixture	Range of BOD	mL	Range of BOD
0.01	20,000 – 70,000	0.02	30,000 – 105,000
0.02	10,000 – 35,000	0.05	12,000 – 42,000
0.05	4,000 – 14,000	0.10	6,000 – 21,000
0.1	2,000 – 7,000	0.20	3,000 – 10,500
0.2	1,000 – 3,500	0.50	1,200 – 4,200
0.5	400 – 1,400	1.0	600 – 2,100
1.0	200 – 700	2.0	300 – 1,050
2.0	100 – 350	5.0	102 – 420
5.0	40 – 140	10.0	60 – 210
10.0	20 – 70	20.0	30 – 105
20.0	10 – 35	50.0	12 – 42
50.0	4 – 14	100	6 – 21
100	0 – 7	300	0 – 7

Source: Sawyer and McCarty (1978, p. 424). Used with permission.

plunger-type mixing rod. Transfer the diluted sample to three BOD bottles. Immediately measure the DO in one bottle; incubate the other two at 20°C for five days.

$$mg/L = \frac{D_1 - D_5}{P}$$

where

D_1 = initial DO of diluted sample
D_5 = 5-day DO of diluted sample
P = decimal fraction of sample used

3. *Chlorinated Samples*: Chlorinated samples are neutralized by sodium sulfite. The appropriate quantity of Na_2SO_3 solution to add to the sample is determined on a 100–1000 mL portion of the sample by adding 10 mL of the 1→50 H_2SO_4 solution, followed by 10 mL of the KI solution and titrating with 0.025 N Na_2SO_3 to the starch-iodide endpoint. Add to a volume of sample the quantity of Na_2SO_3 to the starch-iodide endpoint. Add to a volume of sample the quantity of Na_2SO_3 determined above to neutralize the chlorine. Test an aliquot of neutralized sample by the above method to check for residual chlorine. Use the neutralized sample for the BOD test.
4. *Alkaline or Acidic Samples*: Adjust using a pH meter to about pH 7.0 with 1 N H_2SO_4 or 1 N NaOH.

5. *Seeding*: This procedure is used on neutralized chlorine samples and the samples which need a biological population capable of oxidizing the organic matter in the wastewater. Use 2 mL seed per liter of diluted sample (0.2% seed). A seeded blank must be run with the samples. A seed correction factor, f, is determined by setting up a separate series of seed dilutions using varying proportions of seed and dilution water. Choose the one dilution that gives a 40–70% oxygen depletion in 5 days. Designate that seed dilution as B_1 and B_5 for initial DO and for 5-day DO.

$$\text{mg BOD}_5/\text{L} = \frac{\overbrace{(D_1 - D_5)}^{\text{sample}} - \overbrace{(B_1 - B_5)f}^{\text{seed blank}}}{P}$$

where

D_1 = initial DO of diluted sample
D_5 = 5-day DO of diluted sample
B_1 = initial DO of seeded blank dilution that gives 40–70% oxygen depletion in 5 days
B_5 = 5-day DO of seeded blank dilution that gives 40–70% oxygen depletion in five days
$f = \dfrac{\text{\% seed in diluted sample (should be 0.2\%)}}{\text{\% seed in seed blank}}$
P = decimal fraction of sample used

5.2 Chemical Oxygen Demand (COD) Manual Method: Dichromate Reflux

A. General

1. References: See *Standard Methods* (1989, pp. 5-10 through 5-14), and EPA (1983, Method 410).
2. Outline of Method:
   The chemical oxygen demand (COD) determination provides a measure of the oxygen equivalent of that portion of the organic matter in a sample that is susceptible to oxidation by a strong chemical oxidant. The dichromate reflux method has been selected for the COD determination because it has advantages over other oxidants in oxidizability, applicability to a wide variety of samples, and ease of manipulation. Most types of organic matter are destroyed by boiling a mixture of chromic and sulfuric acids. A sample is refluxed with known amounts of potassium dichromate and sulfuric acid, and the excess dichromate is titrated with ferrous ammonium sulfate. The amount of oxidizable organic matter, measured as oxygen equivalent, is proportional to the potassium dichromate consumed. This technique is designed for typical sewage, but there are many modifications possible for other samples; see *Standard Methods* (1989, pp. 5-10 through 5-14) for details.
3. *Note*: Mercury is a hazardous waste. All mercury-containing waste solutions must be stored in labeled containers.

B. Special Reagents

Always use high-quality laboratory DDW.

1. *Mercuric Sulfate:* Use analytical grade to complex chlorides and remove them from the reaction.
2. *Ferroin Indicator Solution*: Dissolve 1.485 g 1,10-phenanthroline monohydrate together with 0.695 g $FeSO_4 \cdot 7H_2O$ in DDW and dilute to 100 mL.
3. *Concentrated Potassium Dichromate Solution, 0.25 N*: Dry approximately 15 g of $K_2Cr_2O_7$ (primary standard grade) at 103°C for 2 hr. Dissolve exactly 12.259 g of the specially dried orange crystal in DDW

and dilute to exactly 1 L. Nitrite-nitrogen exerts a COD of 1.14 mg per mg NO_2-N. Add 0.12 g of sulfamic acid per liter to the concentrated dichromate solution to eliminate the interference of nitrites in concentrations up to 6 mg NO_2-N/L in the sample.

4. *Concentrated Sulfuric Acid Catalyst*: Add 22 g of silver sulfate (Ag_2SO_4) to a full 9 lb bottle of conc. H_2SO_4. Allow 1–2 days for dissolution with stirring and then attach the bottle to an acid repipet. Set the repipet to deliver 30.0 mL. Silver sulfate catalyst is used to more effectively oxidize straight chain organic compounds.
5. *Ferrous Ammonium Sulfate (FAS) Solution approximately 0.25 N*: Dissolve 98 g of $Fe(NH_4)_2(SO_4)_2 \cdot 6H_2O$ in approximately 600 mL DDW. Add 20 mL of conc H_2SO_4. Cool. Dilute to 1 L with DDW.
6. *Dilute FAS Titrant, approximately 0.10 N*: Dilute 400 mL of the 0.25 N FAS solution to 1 L with DDW. Standardize daily.
7. *Chromic Acid Glassware Cleaning Solution*: Use extreme caution. Wear safety glasses, gloves, and protective clothing while making and using this solution. Place 75 mL of DDW in a 250 mL beaker. Add sodium dichromate ($Na_2Cr_2O_7$) while stirring until the solution is saturated (a few crystals will remain on the bottom). Decant this solution into a 3 or 4 L beaker or Erlenmeyer flask. Set it up on a magnetic stirrer. Slowly and carefully add 2.2 L (one 9 lb bottle) of concentrated sulfuric acid (H_2SO_4) to the saturated $Na_2Cr_2O_7$ solution. Stir carefully to dissolve.
8. *50% v/v H_2SO_4 Glassware Cleaning Solution*: In a hood, carefully add 500 mL conc H_2SO_4 to 450 mL of DDW. Cool and dilute to 1 L.
   a. *Note*: Care and cleaning of glassware is of extreme importance in the COD determination, since a very small amount of organic contamination will drastically change the results. Flasks and condensers should be rinsed carefully with 50% H_2SO_4, then DDW, prior to each use.
   b. *Glass Beads*: Glass beads may be used repeatedly if they are soaked in the 50% H_2SO_4 cleaning solution, rinsed with DDW, and air dried in the 103°C oven after each use. Cleaned beads should be stored in a covered bottle.
   c. *Chromic Acid Cleaning*: When heavy deposits build up on the reflux flasks, they should be treated with chromic acid cleaning solution. Add about 100 mL of the cleaning solution to each flask; let this stand 1 or 2 days. Pour the cleaning solution back into its bottle for reuse.

C. Standardization

Standardization must accompany each set of determinations.

1. *Standardization of FAS Titrant*:

   Note: This can be completed while the samples and blanks are refluxing. See 5.2 D.

   Dilute 400 mL of the 0.25 N FAS solution to 1 L with fresh DDW. This produces approximately 0.10 N FAS titrant. Rinse and fill the buret with the titrant.

   Pipet 10.0 mL of 0.25 N potassium dichromate solution into a clean flask (see 5.2 B) and dilute to approximately 100 mL with fresh DDW. Add 30 mL of conc H_2SO_4 catalyst, swirling while adding. Cool. Add 3–4 drops of Ferroin Indicator, and titrate to the red endpoint (blue-green to reddish brown) with dilute FAS solution. Calculate the FAS normality.

$$\text{FAS Normality} = \frac{(\text{mL } K_2Cr_2O_7)\ (\text{N } K_2Cr_2O_7)}{\text{mL FAS used}}$$

$$= \frac{(10)\ (0.25)}{\text{mL FAS used}} \approx 0.10$$

2. *Deionized Water Blanks*: Prepare three DDW blanks for each set of samples. Proceed as described below using 20 mL of fresh DDW instead of 20 mL of sample.
3. *KHP Standard*: Potassium acid phthalate (KHP; potassium biphthalate) has a theoretical COD of 1.76 g/g. A 98–100% recovery of the theoretical oxygen demand can be expected with this compound.

   Prepare a KHP standard solution whose theoretical COD approximates the COD level of the samples. For example, dissolving 0.4252 g KHP (dried at 103°C) in fresh DDW and diluting to 1 L will produce a 500 mg COD/L solution. Proceed as described below using 20 mL of KHP solution instead of 20 mL of sample. The COD value determined for this solution is used as a procedural check. If the calculated COD value (see 5.2 E) varies by more than ± 10% from the theoretical value, there is a good possibility the FAS titrant and the $K_2Cr_2O_7$ solutions are bad and should be remade.

D. Procedure

1. Rinse 300 mL reflux flasks with 50% H_2SO_4 cleaning solution and DDW.
2. Place 4 or 5 cleaned, dried glass beads in each of the flasks.
3. Add 1–2 scoops (approximately 0.4 g) of mercuric sulfate to each flask. Such an addition will complex any chloride ions present (up to 40 mg Cl^-) and prevent their interference. Add a 20.0 mL sample (or an aliquot diluted to 20.0 mL with DDW) and mix.
4. Add exactly 10 mL of 0.25 N $K_2Cr_2O_7$. Swirl and slowly add 30 mL of the conc H_2SO_4 reagent, swirling the solution throughout the acid addition. The reflux mixture must be thoroughly mixed before the heat is applied. Acid volume : sample + dichromate volume ratio must be 1:1.
5. Attach condensers to all flasks making sure that all connections are secure. Be absolutely sure that cold water is circulating through all of the water jackets surrounding the condensers. Turn on the hot plate to the highest temperature setting. When the solutions begin to boil, begin timing and reflux for exactly 2 hr. Fumes will be seen condensing inside the columns, but there should never be acid fumes escaping from the tops of the condensers.
6. After the 2 hr reflux period, turn off the heat and allow the samples to cool to room temperature, leaving the flasks and the condensers connected.

   Note: If any of the samples are green in color, this indicates that all of the dichromate oxidant has been used (reduced). A dilution of the sample must be made and the reflux procedure repeated.

7. Wash down the condensers with 40 mL of DDW and remove the flasks from the condensers. Dilute the mixture with 50 mL DDW and cool to room temperature.
8. Add 3 drops of Ferroin indicator and titrate with the standard dilute FAS (approximately 0.10 N solution). The endpoint is indicated by a blue-green to reddish brown color change.

   Caution: Ferroin indicator is destroyed when added to hot solution.

E. Calculations

$$\text{mg COD/L} = \frac{(A - B)\ N\ (8000)}{\text{mL sample}}$$

where

A = mL FAS used for sample
B = mL FAS used for blank
N = normality of FAS

If 10 mL 0.25 N $K_2Cr_2O_7$ is used for titrant standardization and 20 mL sample volumes are used:

$$\frac{N(8000)}{\text{mL sample}} = \frac{(\text{mL } K_2Cr_2O_7)(\text{N } K_2Cr_2O_7)(8000)}{(\text{mL sample})\ (\text{mL FAS used to standardize})}$$

$$= \frac{(10)\ (0.25)\ (8000)}{(20)\ (\text{mL FAS used to standardize})}$$

$$= \frac{(2.5)\ (400)}{(\text{mL FAS used to standardize})}$$

$$= \frac{1000}{(\text{mL FAS used to standardize})}$$

Therefore,

$$\text{mg COD/L} = \frac{1000\ (A - B)}{(\text{mL FAS used to standardize})}$$

F. Notes

The procedure is applicable for 25–900 mg COD/L. If samples are expected to exceed this range, appropriate dilutions must be made. If low levels of COD are expected (generally 5–100 mg COD/L), use diluted solutions of $K_2Cr_2O_7$ and FAS (i.e. 0.025 N $K_2Cr_2O_7$ titrated with 0.01 N FAS).

5.3 Chemical Oxygen Demand (COD) Ampule Method

A. General

1. References: See *Federal Register* (Vol. 43, No. 45, March 7, 1978) and *Standard Methods* (1989, pp. 5-10 through 5-11; 5-15 through 5-16).
2. Outline of Method:

   Chemical Oxygen Demand (COD) determines the quantity of oxygen required for oxidation of the inorganic and organic matter in a water sample under controlled conditions of oxidizing agents, temperature, and time. Dichromate is used as the oxidizing agent here as in the manual method. However, the manual digestion technique is modified to use semi-micro volumes of samples and reagents in ampules with reflux digestion performed in an autoclave or an oven.

   Prepackaged Reagents: Prepared ampules, twist tubes, or vials may be purchased from O. I. Inc., College Station, Texas, or Hach Chemical, Loveland, Colorado, containing a premixed digestion solution and catalyst suitable for the determination of various COD concentrations.

   If premixed COD ampules or vials have been obtained, preparation of the digestion solution and the silver sulfate catalyst solution is not required, although reagent preparation is presented in steps 5.3 B.1 through B.4 for those who wish to prepare the solutions and glassware in their laboratory.

   It should be noted that premixed ampules or vials should be stored in light-proof containers at room temperature (70°C). Also be aware of the guaranteed shelf life of these products and do not use them after the expiration date.

   The amount of oxidizable organic matter, measured as oxygen equivalent, is proportional to the dichromate oxidant consumed after digestion. For low level COD, the amount of unconsumed dichromate (as Cr^{+7}, orange in color) is measured spectrophotometrically. For high level COD, the amount of reduced oxidant (as Cr^{+3}, green in color) is measured spectrophotometrically.
3. *Note*: Mercury is a hazardous waste. All spent ampules and waste solutions containing mercury must be stored in labeled waste containers and disposed of properly.

B. Special Reagents and Materials

1. *Ampules*: Wrap the ampule or vial openings with aluminum foil and fire for 30 min at 550°C; cool overnight. Do not rinse the ampules with 20% H_2SO_4.
2. *Glassware Wash—20% H_2SO_4*: In a hood, carefully add 200 mL conc H_2SO_4 to 600 mL DDW. Cool and dilute to 1 L.
3. *Digestion Solution*:
   a. High level
   Mix 10.216 g $K_2Cr_2O_7$ (primary std; dried), 333 g mercuric sulfate, and 167 mL conc H_2SO_4. Carefully add this solution to 600 mL DDW. Cool. Dilute to 1 L.

   Potassium Dichromate Normality = 0.2084

   b. Low Level
   Mix 2.554 g $K_2Cr_2O_7$ (primary std; dried), 33.3 g mercuric sulfate and 167 mL conc H_2SO_4. Carefully add this solution to 600 mL DDW. Cool. Dilute to 1 L.

   Potassium Dichromate Normality = 0.0521

4. *Silver Sulfate Catalyst Solution (COD acid)*: Carefully add 22 g Ag_2SO_4 to a 9 lb bottle conc H_2SO_4 (2.2 L). Stir until dissolved.
5. *Potassium Hydrogen Phthalate Stock Solution*:
   a. High Level (25–900 mg COD/L)
   Dry 10–12 g potassium hydrogen phthalate (KHP) in a 120°C oven for at least 1 hr. Dissolve 8.500 g KHP in 800 mL DDW and dilute to 1 L. This stock solution is 10,000 mg COD/L.
   b. Low Level (5–100 mg COD/L)
   Dry 1–2 g potassium hydrogen phthalate (KHP) in a 120°C oven for at least 1 hr. Dissolve 0.8500 g in 800 mL and dilute to 1 L. This stock solution is 1000 mg COD/L.

C. Standardization

Note: Wash all glassware and pipets with 20% H_2SO_4. Rinse well with DDW.

1. High Level: Use the 10,000 mg COD/L stock solution

Final Concentration mg COD/L	Amount Stock Solution (10,000 mg/L)
100	1.0 mL diluted up to 100 mL
500	5.0 mL diluted up to 100 mL
700	7.0 mL diluted up to 100 mL
900	9.0 mL diluted up to 100 mL
25	Use 5.0 mL of 500 mg/L standard and dilute up to 100 mL

Note: In the upper ranges of concentrations, the standard will be somewhat green in color.

2. Low Level: Use the 1000 mg COD/L stock solution

Final Concentration mg COD/L	Amount Stock Solution (1000 mg/L)
5	1.0 mL diluted up to 200 mL
10	1.0 mL diluted up to 100 mL
25	5.0 mL diluted up to 200 mL
50	5.0 mL diluted up to 100 mL
100	10.0 mL diluted up to 100 mL
200	20.0 mL diluted up to 100 mL

Set up duplicates for all standards, and set up three DDW blanks.

Note: Also set up a 200 mg/L "standard" for use later to set the zero on the spectrophotometer (see 5.3 D 8).

D. Procedure (For Premixed Vials)

Note: Set up all standards and samples in duplicate.

1. Preheat the laboratory COD oven to 150°C or prepare the autoclave (121°C, 15–17 psi) for a 2-hr cycle. A microwave procedure has also been reported (Jardim and Rohwedder, 1989).
2. Remove the cap from the premixed COD vial and carefully add 2.5 mL of the sample (or standard solution) to the vial such that it forms a layer on the top. *Do not mix at this point.*
3. Replace the cap on the vial and tighten. Caps may be broken by overtightening, but they need to be suitably tight to prevent leakage.

4. Mix the vials carefully. Use insulated gloves and wear eye protection. Discard any vials that leak and prepare new replacement samples (or standards).
5. Prepare standards and blanks just as the samples were prepared.
6. Heat the vials at 150°C for 2 hr or autoclave for 2 hr.
7. Cool the vials and protect them from light. Mix well. Allow 10 min for any suspension to settle.
8. Transfer a portion of the liquid from the vial to a spectrophotometer cuvette and read the absorbance.

   For high level COD concentrations, set the spectrophotometer at 600 nm and zero the instrument using a reagent blank (DDW and reagents). Read the absorbance of the digested standards and samples at 600 nm.

   For low level COD concentrations, an inverse curve is used, i.e., the reagent blank will have the highest absorbance. Set the spectrophotometer at 400 nm and zero the instrument using a standard higher than the upper limit of the concentration range (i.e., 200 mg/L, etc.). Read the absorbance of the digested standards and samples at 440 nm.

E. Calculation

Calculate the COD of the sample as follows:

$$\text{COD(mg/L)} = (\text{Abs}_{\text{sample}})(m^{-1})(\text{df}) + B$$

where

$$m^{-1} = \frac{\Delta \text{Concentration}}{\Delta \text{Absorbance}} = \frac{C_2 - C_1}{\text{Abs}_2 - \text{Abs}_1}$$

C_2 = COD (mg/L) of any standard
C_1 = COD (mg/L) of any standard lower than C_2
Abs_2 = The absorbance of standard C_2
Abs_1 = The absorbance of standard C_1
df = dilution factor

If a 10% sample is used, df = 10; if a 5% sample is used, df = 20.

For high level,
B = zero for the high level procedure; and

for low level,

B = COD (mg/L) of the highest standard for the low level procedure.

Alternately, calculate the concentration of COD in a sample using a linear regression of the calibration standards, or plot the absorbance of the calibration standards against the calibration concentrations and compute the sample concentration directly from the linear standard curve. If dilutions were made, multiply by an appropriate dilution factor.

5.4 Oil and Grease
Partition-Gravimetric Method

A. General

1. References: See *Standard Methods* (1989, pp. 5–41 through 5–44) and EPA (1983, Method 413.3).
2. Outline of Method:
   Liquid-liquid extraction is the method most widely employed in separating organic compounds from aqueous mixtures in which they are found or produced. This procedure involves the distribution of a solute between two immiscible solvents. The oils and greases are extracted from the aqueous solution by direct contact with an immiscible organic solvent, trichlorotrifluoroethane. The organic solvent is then separated from the aqueous phase, dried, and evaporated to determine the extractable residue by gravimetric techniques. This method is not applicable to light hydrocarbons that volatilize below 70°C (i.e., petroleum fuels from gasoline through #2 fuel oils). Some crude oils and heavy fuel oils contain nonextractable residues so the recoveries will be low.

   Note: Samples should be collected in glass containers only and preserved with acid.

B. Special Reagents and Apparatus

1. *Sulfuric Acid, 1 + 1 Solution*: In a hood, carefully mix a volume of concentrated sulfuric acid with an equal volume of DDW. Cool.
2. *Organic Solvent*: Use trichlorotrifluoroethane (Freon or Fluorocarbon-113), boiling point, 47°C.
3. *Sodium Sulfate, Anhydrous*: Use analytical reagent grade Na_2SO_4.
4. *Separatory Funnel, Glass Only*: Use a 2-liter separatory funnel (sufficient capacity for 1 L of sample plus the addition of acid and solvent while still leaving space for proper agitation) with a Teflon stopcock.
5. Use *250 mL Boiling Flasks and Distillation Columns.*
6. *Water or Steam Bath, Heating Mantle or Hot Plate*: It should be capable of maintaining 70°C.
7. *Filter Paper*: Whatman No. 40 or equivalent, 11 cm diameter (glass wool may also be substituted).
8. Use *Boiling Chips.*

C. Standardization

None is required. However, as a check on technique, it is a good idea to run known samples of vegetable or fuel oil through the procedure occasionally. Percent recover may vary from 80–95%.

The method is applicable in the range of 5 to 1000 mg/L of extractable matter.

Always set up two blanks using 1 L of DDW to run through the procedure listed below.

D. Procedure

1. Rinse all glassware with trichlorotrifluoroethane to remove any traces of grease and oil. Save all solvent for repurification later. Predry boiling flasks in a 103°C oven, cool, and store in a desiccator. Add a boiling chip to each and record initial weights.
2. Collect about 1 L of sample and mark the bottle at the water miniscus for later determination of sample volume. It is best to analyze the entire sample collected. Grease and oil may float to the top and removal of a portion of the sample will not result in an accurate determination.
3. If the sample was not acidified at the time of collection, add 5 mL of 1 + 1 H_2SO_4 to the sample. Mix well. Check the pH by touching pH paper to the cap liner. The pH should be 2 or lower. Add more acid if necessary and record the volume added.
4. Transfer the sample to a 2-liter separatory funnel. Rinse the sample bottle with 15 mL of trichlorotrifluorethane and pour the washings into the separatory funnel. Rinse the sample bottle with an additional 25 mL of solvent and pour into the separatory funnel also.
5. Shake the separatory funnel vigorously for 2 min.

   Caution: Vent the excess pressure after the first few shakes by holding the funnel at an angle and opening the stopcock momentarily.

   Note: Some samples will tend to form a stable emulsion evidenced by a somewhat indistinct third layer in the separatory funnel. If this is the case, shake the separatory funnel gently for 5–10 min.

6. Allow about 10 min for the layers to separate. Drain the organic solvent layer into a clean, dry 250 mL Erlenmeyer flask that contains 0.5 to 1 g Na_2SO_4 drying agent. This flask does not need to be preweighed. Swirl the solvent. If the solvent appears cloudy or if water is collecting around the Na_2SO_4 crystals, then add another 0.5 g drying agent. When the

solution is clear and the crystals do not appear wetted with water, pour the solvent solution through a funnel containing solvent-moistened filter paper into a clean, tared distilling flask. If a clear solvent layer cannot be obtained in the funnel or if there is evidence of an emulsion, add 1 g Na_2SO_4 drying agent to the filter paper cone and slowly drain the solvent onto the crystals. If the solvent is still cloudy, more drying agent may be added 1 g at a time.

7. Repeat the rinsing and extraction procedures two additional times on the sample remaining in the separatory funnel. Each time add the organic solvent layer to the boiling flask containing the first solvent extract.
8. Rinse the tip of the separatory funnel, the filter paper, and then the funnel with a total of 10–20 mL of solvent, and collect these rinsings in the boilings flask also. Make sure that the solvent is clear and free of water.
9. Connect the distilling flask to a distillation column and immerse the flask in a water or steam bath or place on a heating mantle or hot plate.
10. Evaporate the solvent almost to dryness. The water bath or heating mantle should be maintained at 70°C. The temperature above the boiling solvent at the inlet to the distillation column should be monitored also. Here the temperature should be about 47°C (~45°C at high altitude) to assure recovery of pure Freon solvent. Collect the trichlorotrifluoroethane for repurification and reuse. Cool the flask. Volatilize the last portions of the solvent with a gentle stream of nitrogen gas or air blown directly into the flask.

    Caution: Laboratory air supplies often contain grease and oil; install a trap in the line if air is used.

11. Wipe the outside of the boiling flask carefully to remove moisture and fingerprints.
12. Place in a desiccator for 30 min to 1 hr. Reweigh to determine the amount of extractable residue collected.
13. Place samples in the desiccator again for 30 min to 1 hr. Weigh again to determine if all samples have been dried to constant weight. Repeat desiccating procedures until weight is constant.
14. Refill the sample bottles to the marked line with water and transfer to a graduated cylinder to determine the sample volume in liters. Correct this volume for acid addition.

E. Calculations

$$\text{mg extractable organic matter/L} = \left[\frac{(A - B) - C}{D}\right] \times 1000$$

where

- A = total weight (flask + boiling chips + residue) in grams
- B = tare weight (flask + boiling chips)
- C = average blank determination, residue from equivalent volume of DDW (total weight – tare weight).
- D = volume of sample in liters, corrected for acid addition if necessary

F. Notes

1. Repurify the solvent by boiling in a distillation apparatus and collecting the solvent at its boiling point (47°C).
2. Sometimes the glassware used cannot be cleaned adequately with a Freon rinse only. If visible traces of grease and oil remain, place the glassware in a muffle furnace at 550°C for 30 to 60 min to volatilize the residues.

5.5 Surfactants
Methylene Blue Active Substances (MBAS) Method

A. General

1. References: See *Standard Methods* (1989, pp. 5-55 through 5-56; 5-59 through 5-63), and EPA (1983, Method 425.1).
2. Outline of Method:
   Linear alkylate sulfonate (LAS) is a somewhat biodegradable surfactant found in detergents. In this procedure, LAS and other anionic surfactants react with methylene blue, forming a blue salt. The salt is soluble in chloroform; the color intensity that is produced is proportional to the concentration of the anionic surfactants.
3. Organic sulfates, sulfonates, carboxylates, phosphates, and phenols, and inorganic cyanates, chlorides, nitrates, and thiocyanates can cause interferences. See *Standard Methods* (1989, pp. 5-60 through 5-61) for a discussion of interferences.

B. Special Reagents and Equipment

1. *Stock Linear Alkylate Sulfonate (LAS) Solution*: Weigh an amount of reference material equal to 1.000 g LAS on a 100% active basis. To obtain this, divide 1.000 g by the percent active stated on the ampule label. For example, if the LAS is stated as 5.69% active, divide:

   $$\frac{1.000 \times 100}{5.69} = 17.575 \text{ g of LAS solution to be weighed}$$

   Dissolve in DDW and dilute to 1 L. Store in the refrigerator. Stock solution is 1000 mg LAS/L.
2. *Standard LAS Solution*: Dilute 10.0 mL stock LAS solution to 1 L with DDW. Prepare this daily. The standard solution is 10 mg LAS/L.
3. *Phenolphthalein Indicator Solution*: Dissolve 5 g phenolphthalein disodium salt in 500 mL of 95% ethyl alcohol and add 500 mL DDW.
4. *Sodium Hydroxide, 1 N*: With stirring, dissolve 40 g NaOH in DDW. Cool and dilute to 1 L.
5. *Sulfuric Acid, 1 N*: Carefully add 28 mL conc H_2SO_4 to 500 mL DDW. Cool and dilute to 1 L.

6. *Chloroform*: Use reagent grade. (Note: When using chloroform, always work under a hood. Use safety glasses or a mask and wear gloves.)
7. *Methylene Blue Reagent*: Dissolve 100 mg methylene blue in 100 mL DDW. Transfer 30 mL to a 1000 mL flask. Add 500 mL DDW, 6.8 mL conc H_2SO_4, and 50 g mono-sodium dihydrogen phosphate monohydrate, $NaH_2PO_4 \cdot H_2O$. Shake until dissolution is complete. Dilute to 1 L.
8. *Wash Solution*: Add 6.8 mL conc H_2SO_4 to 500 mL DDW in a 1000 mL flask. Then add 50 g $NaH_2PO_4 \cdot H_2O$ and shake until dissolution is complete. Dilute to 1 L with DDW.
9. *Separatory Funnels*: Use 500 mL separatory funnels with Teflon stopcocks.

C. Standardization

1. Prerinse four separatory funnels with a small amount of chloroform to remove any existing LAS residue. (Note: When using chloroform, always work under a hood. Use safety glasses or a mask and wear gloves.)
2. Prepare a standard curve by adding 0, 1.0, 5.0, and 9.0 mL of the standard LAS solution to four separatory funnels. Add DDW to make the total volume of water in each separatory funnel (about 100 mL).

Final Concentration μg LAS/L	mL Standard LAS
0	0
100	1.0
500	5.0
900	9.0

Follow the procedure below (extraction, color development, and color measurement) for each of the standards.

D. Procedure

1. Volume of sample: Select the volume of the water sample to be tested based on the expected LAS concentration:

Expected LAS Concentration mg/L	Sample Volume to be Used, mL
0.025– 0.080	400
0.08 - 0.40	250
0.4 - 2.0	100
2 - 10	20.0
10 -100	2.00

If a sample of less than 100 mL is indicated, dilute to 100 mL with distilled water; if 100 mL or more are used, extract the entire sample.

2. Prerinse the separatory funnels with a small amount of chloroform to remove any existing LAS.

   Note: When using chloroform, always work under a hood. Use safety glasses and/or a mask, and wear gloves.

3. Add the sample solution to the separatory funnel. Make the solution alkaline by dropwise addition of 1 N NaOH, using 10 drops phenolphthalein indicator. Discharge the pink color by dropwise addition of 1 N H_2SO_4.
4. Add 10 mL chloroform and 25 mL methylene blue reagent. Rock the funnel vigorously for 30 seconds and let the phases separate. Be sure to vent any excess pressure in the funnel by tilting it and opening the stopcock momentarily. Excessive agitation may cause emulsion trouble. Some samples require a longer period of phase separation than others. Before draining the chloroform layer, swirl the sample gently, then let it settle.
5. Draw off the chloroform layer into a second separatory funnel when the layers have separated. Rinse the delivery tube of the first separatory funnel with a small amount of chloroform. Repeat the extraction three times, using 10 mL of chloroform each time. If the blue color in the water phase becomes faint and disappears, discard the sample and repeat the determination, using a smaller sample size.
6. Combine all chloroform extracts in the second separatory funnel. Add 50 mL wash solution and shake vigorously for 30 seconds. Emulsions do not form at this stage. Let this settle, swirl the contents, and then draw off the chloroform layer through glass wool that has been pre-extracted with chloroform into a 100 mL volumetric flask. Extract the wash solution twice with 10 mL chloroform, adding these to the volumetric flask. Rinse the glass wool and the funnel with chloroform. Collect the washing in the volumetric flask and dilute to the mark with chloroform. Place the stopper on the flask and twist slightly to prevent leakage. Mix by inverting several times.
7. Measurement: Determine the absorbance of the samples and standards at 652 nm against a blank of chloroform.

Note: Always save the waste chloroform in this procedure. It can be redistilled and used again.

E. Calculation

Calculate the concentration of MBAS in a sample using a linear regression of the calibration standards, or plot the absorbance of the calibration standards against the calibration concentrations and compute the sample concentration directly from the linear standard curve.

5.6 Organic Carbon in Soils
Dichromate Method

A. General

1. Reference: See Walkley (1935).
2. Outline of Method:
   Soils are exposed to a strong chemical oxidant under acidic conditions and the organic carbon is measured (stoichiometrically as in COD) as oxidant consumed.

B. Special Reagents

1. *Potassium Dichromate, 1 N*: Dry 53 g $K_2Cr_2O_7$ at 103°C for 2 hr. Dissolve exactly 49.036 g of the dried crystal in DDW and dilute to 1 L. Nitrite-nitrogen exerts a COD of 1.14 mg per mg NO_2-N. Add 0.12 g of sulfamic acid per liter of dichromate solution to eliminate the interference of nitrite up to 6 mg/L.
2. *Sulfuric Acid, 0.5 N*: Carefully add 14 mL conc H_2SO_4 to about 400 mL DDW and dilute to 1 L.
3. *Ferrous Sulfate, 1 N*: Dissolve 278.02 g of $FeSO_4 \cdot 7H_20$ in 0.5 N H_2SO_4 and dilute to 1 L using 0.5 N H_2SO_4.
4. *Ferroin Indicator Solution*: See 5.2.
5. *Concentrated H_2SO_4-Ag_2SO_4 Reagent*: See 5.2.
6. *Phosphoric Acid, 85%*: Use analytical reagent grade conc H_3PO_4.

C. Standardization

1. Measure 10.0 mL 1 N $K_2Cr_2O_7$ into a 500 mL Erlenmeyer flask. Add 20 mL conc H_2SO_4; swirl for 1 min. Allow this to stand for 30 min to cool the solution.
2. Add approximately 200 mL DDW, followed by approximately 10 mL of 85% H_3PO_4 and 4 drops of ferroin indicator.
3. Titrate with the ferrous sulfate solution until the color changes from green to red.

   Note 1: The stoichiometry of the reaction has been outlined by Sawyer and McCarty (1978). It is as follows:

$$C_nH_aO_b + cCr_2O_7^{-2} + 8cH^+ \xrightarrow{\Delta} nCO_2 + \frac{8c + a}{2} H_2O + 2cCr^{+3}$$

$$\text{where } c = \frac{2n}{3} + \frac{6}{a} - \frac{b}{3}$$

As can be seen, the amount of chromate reduced per carbon atom oxidized will vary depending upon the source of the organic carbon. For instance, if $C_6H_{12}O_6$ is the carbon source, two chromates will oxidize three carbons; if C_2H_2OH is the carbon source, three chromates will oxidize four carbons. It has been determined that a stoichiometry of two chromates for three carbons is an average relationship for most soils. This corresponds to a –4 change in oxidation state for each carbon and gives a milliequivalent weight for each carbon of 3 mg. Using 1 N $K_2Cr_2O_7$, this would theoretically equal 3 mg carbon oxidized for every mL of $K_2Cr_2O_7$ reduced.

Note 2: The percent recovery of the method is something that is variable from soil to soil and should be checked by a dry combustion of carbon on duplicate samples.

Note 3: Mean recoveries of about 77% were found. This gives a correction factor of 100/77 = 1.3:

$$1 \text{ mL } 1 \text{ N } K_2Cr_2O_7 = 3 \text{ mg C} \times 1.3 = 3.9 \text{ mg C}$$

D. Procedure

1. Air dry soil samples for 3 days. Grind the soil sample in a mortar and pass the sample through a 0.5 mm sieve.
2. Weigh 0.5 to 3.0 g of sample to the nearest mg and place it in a 500 mL Erlenmeyer flask. Choose the sample size so as to reduce 3.0 to 7.0 mL of dichromate.
3. Add 10.0 mL of 1 N $K_2Cr_2O_7$ to the flask and then add 20 mL conc H_2SO_4. Use a quick delivery automatic pipet for the sulfuric acid; add the 20 mL of acid in less than 2 seconds.
4. Swirl the flask for 1 min and then allow it to stand for 30 min.
5. Add approximately 200 mL DDW, approximately 10 mL 85% H_3PO_4, and 4 drops of ferroin indicator.
6. Titrate with the ferrous sulfate solution until the color changes from green to red. The end point should be sharp and determined within a fraction of a drop.

E. Calculation

$$\frac{\text{mg Org C}}{\text{g sample}} = \frac{(A - B)(3.9)}{\text{g sample}}$$

where

A = mL $FeSO_4$ titrant used for standard

B = mL $FeSO_4$ titrant used for sample

g sample = g of air dried, sieved sample

5.7 Preparation of Samples for Trihalomethane Determination Gas Chromatography Method

A. General

1. References: See *Federal Register* (Vol. 44, No. 231, November 29, 1979) and *Standard Methods* (1989, pp. 6–104 through 6–112).
2. Outline of Method:
   Trihalomethanes (THMs) are a group of organic compounds that are defined as containing one carbon atom, one hydrogen atom, and three halogen atoms. The four common trihalomethane compounds in chlorinated drinking water sources are trichloromethane (chloroform), dichlorobromomethane, dibromochloromethane, and tribromomethane (bromoform). Trihalomethanes are analyzed by gas chromatography (electron capture Ni^{63} detector). They usually occur in small quantities (<100 $\mu g/L$); the samples and sampling containers require specific procedures in order to obtain accurate results.

B. Special Reagents and Equipment

1. *L-Ascorbic Acid or Sodium Thiosulfate*: Use powder or crystal, respectively.
2. *Sodium Hypochlorite Solution*: Use reagent grade sodium hypochlorite (NaOCl) 4–6%, or commercial brand Clorox. Sodium hypochlorite readily degrades; check the concentration before use.
3. *Sample Containers*: Use glass containers, 40 mL capacity, with screw caps and Teflon-backed rubber septa cap liners.
4. *Incubator*: This should be capable of maintaining 25°C.
5. *Screw-Cap Vials*: Use 14 mL vials with Teflon tetrafluoroethylene (TFE)-faced silicone septum for liquid-liquid extraction.
6. *Microsyringes*: These should be 10, 25, 100, and 250 mL.
7. *Two 10 mL Glass Hypodermic Syringes with Overlock Tops and Two-Way Valves with Luer ends*: Use Hamilton #86570 or the equivalent.
8. *Temperature-Programmable Gas Chromatograph*: This should have electron capture or other suitable detector.
9. *Chromatographic Column*: Use 4 mm internal dimensions (ID) $\times$ 2 m long glass packed with 3% SP-1000 on Supelcoport 100/120.

10. *Extraction Solvent*: Use halocarbon free hexane (99 Mol% pure, Fisher certified or the equivalent).
11. *Trihalomethane Standards*: These should be 200 mg/mL in methanol, Supelco, or the equivalent.
12. Use *THM-free DDW*.

C. Types of Samples

1. *Instantaneous Total THM*: This is the THM concentration at the time of sampling. In order to obtain this, the sample is preserved immediately with L-ascorbic acid or sodium thiosulfate to prevent further THM formation.
2. *Terminal Total THM*: This is the THM concentration 7 days after the sampling date. In this case, the sample is not treated with a preservative, and it is incubated 7 days.
3. *Maximum Total (THM) Potential*: This is essentially the same as the Terminal Total THM, except that the sample is dosed with chlorine to insure the presence of a chlorine residual after 7 days. Normally, this is determined for samples from unchlorinated sources.
4. *Total Trihalomethanes*: This is the sum in $\mu g/L$ of all trihalomethanes found in a given sample.

D. Procedure

1. Wash all sample containers and cap liners in a detergent approved for chemistry labs. Rinse these thoroughly with DDW.
2. Place the containers and cap liners in a 105°C oven for 1 hr. Allow them to cool in an area free of organics or wrap them in aluminum foil to prevent contamination. Alternately, the glass containers may be placed in a muffle furnace at 550°C for 30 min; cool them in the same manner as described above.

   Note: Do not keep the cap liners in the 105°C oven longer than 1 hr or they may degrade.

3. When containers and cap liners are cool, seal the containers with the cap liners to prevent contamination. Avoid touching any surfaces that will be in contact with the sample water.
4. For Instantaneous THM samples, add 2 to 4 mg L-ascorbic acid or sodium thiosulfate to the empty 40 mL sample containers.
5. When collecting samples, fill the bottles completely, allowing the water

to form an inverted meniscus at the mouth of the container. Do not allow the water to be aerated as the container is filled. Place the rubber liner and lid on, with the Teflon side down. Partially seal the container by twisting the lid on about half way. Gently press on the rubber liner and twist the lid on completely. Check for air by tapping the container and inverting it. If air is present, reseal the container.

6. When sampling from a water tap, allow water to flow for 2 min prior to the collection of the sample. When sampling from an open body of water, obtain a grab sample from a representative area and fill the sample bottle from it.
7. For Maximum Total THM Potential determination, collect the samples in duplicate. Chlorinate by injecting 15 to 20 μL sodium hypochlorite solution per 40 mL sample with a syringe through the cap liner. Incubate the samples at 25°C for 7 days. From the duplicates, check one bottle for a chlorine residual after 7 days. Use the other bottle for the THM analysis.
8. Incubate samples for Terminal Total THM determination at 25°C for 7 days.
9. Store Instantaneous THM samples at 4°C if they cannot be analyzed immediately. Refrigerate (4°C) Terminal and Maximum Potential samples if they cannot be analyzed immediately after the 7-day incubation period.
10. Analyze all samples within 2 weeks of collection.

E. Analysis

1. Set up the gas chromatograph using the following conditions: Injector temperature 100°C and detector temperature 200°C, carrier gas flow rate, 40 mL/min; initial oven temperature, 75°C; initial hold, 1.5; oven temperature program rate, 30°C/min; final oven temperature, 100°C; hold time at 100°C, 2 min.
2. Calibration standards are prepared by injecting microliter quantities of the standard mixture (200 μg/mL) into 100 mL volumetric flasks containing THM-free DDW. Set up the following dilutions for a standard curve.

Final Concentration, μg/L	μL of the 200 μg/mL Standard Solution
2	1 μL injected into 100 mL
10	5 μL injected into 100 mL
50	25 μL injected into 100 mL
100	50 μL injected into 100 mL
300	150 μL injected into 100 mL

After rapidly injecting the standard solution into the flask, place the ground glass stopper on the flask and mix the aqueous solution by inverting the flask three or four times.

3. Remove the plunger from the 10 mL syringe with the attached syringe valve and carefully pour standard solution into the syringe barrel until it overflows. Replace the plunger and adjust to a sample volume of 10.0 mL by venting any air or excess solution. Pipet 2.0 mL of hexane into a clean extraction flask, carefully inject the 10.0 mL contents of the syringe into the extraction flask, seal with the cap containing a TFE-faced septum, and shake vigorously for 1 min. Let this stand for about 1 min until the phases separate. The upper organic phase is taken from the extraction flask and 2.0 μL is injected into the gas chromatograph. Prior to sample analysis, a THM quality control sample should be analyzed to ensure that the procedure and the calibration of the gas chromatograph have been performed properly. If the quality control sample is within the 95% confidence interval, proceed with the sample analysis using procedures identical to those used in analyzing the standard solutions. Analyze all standard solutions and samples in triplicate.

E. Calculation

Based upon the retention times, locate each individual THM in the standard solution on the chromatogram. If the gas chromatograph does not calculate the response of the standards vs concentration, graph either the peak height or peak area vs concentration for the standards.

$$\text{THM}, \mu g/L = PH_{sample}\,(m_H^{-1})$$

or

$$\text{THM}, \mu g/L = PA_{sample}\,(m_A^{-1})$$

where

$$m_H^{-1} = \frac{\Delta \text{ concentration}}{\Delta \text{ Peak Height}} = \frac{C_2 - C_1}{PH_2 - PH_1}$$

$$m_A^{-1} = \frac{\Delta \text{ concentration}}{\Delta \text{ Peak Area}} = \frac{C_2 - C_1}{PA_2 - PA_1}$$

C_2 = THM (μg/L) of any standard
C_1 = THM (μg/L) of any standard lower than C_2

PH_2 = Peak Height of Standard C_2
PH_1 = Peak Height of Standard C_1

PA_2 = Peak Area of Standard C_2
PA_1 = Peak Area of Standard C_1

Calculate total trihalomethane concentration as

$$TTHM = CHCl_3 + CHBrCl_2 + CHBr_2Cl + CHBr_3$$

where all concentrations are in μg/L.

Methods for Biological Determination

6. Methods for Biological Determination

6.1 Plankton and Periphyton Pigments
Chlorophyll
Spectrophotometric Method

A. General

1. References: *Standard Methods* (1989, pp. 10-3 through 10-15; pp. 10-31 through 10-34; pp. 10-48 through 10-54).
2. Outline of Method:
   Three chlorophyll pigments are commonly found in algae: chlorophylls *a*, *b*, and *c*. In this procedure, the chlorophyll is extracted in acetone. The optical density (absorbance) of the extract is determined spectrophotometrically at four different wavelengths. The optical density reading at 750 nm is used as a turbidity correction. Chlorophylls *a*, *b*, and *c* are calculated from the optical density measurements at the other wavelengths. Chlorophyll *a* may be overestimated if pheophytin *a*, which fluoresces near the same wavelength, is present. Pheophytin *a* is a common degradation product of chlorophyll *a*; measurements for both pigments must be made in order to correct for the pheophytin concentration. The ratio of chlorophyll *a* to pheophytin *a* is a good indicator of the physiological condition of the phytoplankton. In the sample, acidification with dilute acid releases the magnesium atom from the chlorophyll *a* molecule, converting it to pheophytin *a*. The fluorescence is read before and after acidification and concentrations for both chlorophyll *a* and pheophytin *a* are calculated.
3. Extract the samples immediately after they are taken. If this is not possible, the samples must be concentrated, frozen, and kept in the dark. Store in this manner for no longer than 30 days.

   Use this method only when chlorophyll concentrations in the extract are high (mg/L range). For a more sensitive technique, refer to the fluorometric method.

B. Special Reagents and Equipment

1. *Acetone, 100%*: Use analytical reagent grade or redistilled.

Note: Acetone is a hazardous waste and cannot be poured down the drain. Waste acetone should be poured into a labeled container and repurified by distillation (boiling point = 56°C).

2. *Acetone, 90% v/v Aqueous Solution*: To 1800 mL of reagent grade or redistilled acetone in a glass bottle, add 200 mL DDW and mix well.

   Note: The optical density at 750 nm is very sensitive to acetone/water proportions. Make up this reagent carefully.

3. *Magnesium Carbonate Suspension*: Add 1.0 g finely powdered $MgCO_3$ to 100 mL DDW.
4. *Hydrochloric Acid, 1N*: Add 5 mL of concentrated HCl to 55 mL of DDW. Store in a dropper bottle.
5. *Tissue Grinder*: Use a Teflon® or glass grinder, preferably one having a grinding tube and pestle with rounded bottoms.
6. *Centrifuge and Centrifuge Tubes*: Use 15 mL graduated tubes with screw caps, chemically resistant to acetone.
7. *Glass Fiber Filters*: Use Whatman GF/C or the equivalent.
8. *Sonicator*: This should be equipped with a probe to dismember and homogenize sample.

C. Standardization

Standardization is not necessary, but it may be useful to run a quality control or chlorophyll *a* standard through the procedure. (See Fluorometric Chlorophyll procedure).

Note: These samples must be stored in the freezer. Even under these conditions, they are only stable for a limited period of time.

D. Procedure

Note: All glassware and cuvettes should be clean and acid-free.

Plankton Samples: These are free-floating or suspended phytoplankton and/or zooplankton.

1. Collect water samples using Van Dorn or Kemmerer water samplers. (See *Standard Methods*, 1989, pp. 10-4 through 10-15 for further discussion). If the water is oligotrophic or phytoplankton densities are expected to be low, collect 6 L of sample. Eutrophic waters will require a volume of 1 to 2 L. Record the depth at which the sample was taken.

2. Filter a known quantity of water through a glass fiber filter to concentrate the sample. Add 0.2 mL magnesium carbonate suspension in the final phase of filtering as a preservative. Store these concentrated samples frozen in a desiccator in the dark if acetone extraction is not possible immediately.
3. Place the filter in a tissue grinder tube and cover it with 2 to 3 mL of 90% acetone. Macerate well without losing any sample.
4. Transfer the filter and solution to a centrifuge tube. Using 90% acetone, rinse the grinding tube and pestle into the centrifuge tube. Adjust the total volume to 5.0 mL (or some other known volume up to 10 mL) with 90% acetone. Allow the solution to stand overnight at 4°C in the dark.
5. Clarify the extract by centrifuging in closed tubes for 20 minutes at 500 g. Pour the supernatant into a calibrated centrifuge tube. Measure and record the total volume of the extract. Take care to keep the supernatant extract as clear as possible and free from suspended pieces of filter paper. Wrap the tubes with aluminum foil to protect them from light.
6. Proceed to step 6 below.

Periphyton Samples: Attached or somewhat motile microscopic organisms associated with surfaces of submerged objects.

1. Scrape samples from the natural or artificial substrates in the body of water. (See *Standard Methods*, 1989, pp. 10-48 through 10-51 for further discussion). Add DDW to the sample bottle to bring it up to a known total volume (usually 100 mL). (It is helpful to calibrate the sample bottle at the 100 mL mark prior to sampling). Record the surface area of the region scraped in in^2 or m^2 and the final total volume. Add 0.2 mL of magnesium carbonate suspension as a preservative. Store these samples frozen in the dark if acetone extraction is not possible immediately.
2. Homogenize the sample by sonication or other suitable means.
3. Withdraw a sample aliquot of known volume to another container. Mix well with an appropriate volume of 100% (not 90%) acetone to make the final solution 90% acetone (example—10 mL of homogenized sample plus 90 mL of acetone).

   Note: If the sample is difficult to homogenize, it may be necessary to perform replicate extractions or increase the size of the sample aliquot. If a large quantity of silt or nonchlorophyll material has been collected with the sample, the periphyton can be centrifuged and the supernatant extracted.

4. Allow the solution to stand overnight at 4°C in the dark.

5. Clarify the extract by centrifuging (as above). Alternately, withdraw the sample carefully from the supernatant to perform relative fluorescence measurements. Avoid agitation of settled material in the extract. Keep the sample bottle or centrifuge tube covered with aluminum foil to minimize light degradation.
6. Measure the optical density at 750, 664, 647, and 630 nm, using a 1 cm cell.

   Note 1: Zero the spectrophotometer using 90% acetone.

   Note 2: With each wavelength change, the spectrophotometer must be zeroed again.

   Note 3: For best results, the absorbance at 664 nm should be greater than 0.2 and less than 1.0 units. Choose the appropriate cell length and/or dilution to do this. Correct the reading back to a 1 cm path length.

   Note 4: Take whatever steps are necessary to maintain a small turbidity correction (optical density at 750 nm).

7. Acidify the extract in the 1 cm cell with 2 drops of 1N HCl. (If a larger cell is used, add proportionately more acid.) Gently agitate the extract.
8. Measure the optical density at 750 and 665 nm no sooner than 1 minute or later than 2 minutes after acidification. Treat all samples identically.

   Note: Acidification results in the shifting of the absorption peak from 664 nm to 665 nm.

9. Pour all samples and waste acetone in a labeled container for repurification by distillation.

E. Calculations

1. Chlorophyll *a*, *b*, and *c* in the extract not corrected for pheophytin *a*: Subtract the reading at 750 nm (turbidity correction) from the other optical densities before using them in the calculations.

$$\text{mg Chl } a/\text{L} = 11.85(\text{OD } 664) - 1.54(\text{OD } 647) - 0.08(\text{OD } 630)$$
$$\text{mg Chl } b/\text{L} = 21.03(\text{OD } 647) - 5.43(\text{OD } 664) - 2.66(\text{OD } 630)$$
$$\text{mg Chl } c/\text{L} = 24.52(\text{OD } 630) - 7.60(\text{OD } 647) - 1.67(\text{OD } 664)$$

   where
   OD 664, OD 647, and OD 630 = turbidity-corrected optical densities at the respective wavelengths using a 1 cm cell.

2. Chlorophyll *a* in the extract corrected for pheophytin *a*:

Subtract the readings at 750 nm (turbidity corrections) from the other optical densities before using them in the calculations.

$$\text{mg Chl } a/\text{L} = 26.73(\text{OD } 664_b - \text{OD } 665_a)$$
$$\text{mg Pheo } a/\text{L} = 26.73\ [(1.7)(\text{OD } 665_a) - \text{OD } 664_b]$$

where

$\text{OD } 664_b$ = turbidity-corrected optical density before acidification using a 1 cm cell

$\text{OD } 665_a$ = turbidity-corrected optical density after acidification using a 1 cm cell

3. Pigment concentrations in plankton samples:

$$\text{mg Chl } X/\text{L} = \frac{\text{mg Chl } X/\text{L} \times \text{A}}{\text{B}}$$

where

mg Chl X/L = concentration of chlorophyll a, b, or c (or pheophytin a) in the extract determined in (1) or (2) above,

A = total volume in liters of the extract supernatant after centrifuging, and

B = volume of the sample filtered in liters.

4. Pigment concentrations in periphyton samples:

$$\text{mg Chl } X/\text{m}^2 = \text{mg Chl } X/\text{L} \times 10 \times C/D$$

where

mg Chl X/L = concentration of chlorophyll a, b, or c (or pheophytin a) in the extract determined in (1) or (2) above,

C = total sample volume in the field situation in liters, usually 0.1 L, and

D = surface area in m^2.

Note: $m^2 = in.^2 \div 1550$.

F. Notes

1. The concentration of plant carotenoids may also be estimated from one of the following equations, depending on which algal group predominates the plankton:

C_{nac} = 4.0(OD 480), if the crop is predominantly Chlorophyta or Cyanophyta.

C_{nac} = 10.0(OD 480), if the crop is predominantly Crysophyta or Pyrrophyta.

where

C_{nac} = the concentration of the nonastacin type (algal) carotenoids in mS.P.U./L (milli-Specified Pigment units).

OD 480 = the optical density at 480 nm corrected for turbidity (OD 750).

6.2 Plankton and Periphyton Pigments
Chlorophyll
Fluorometric Method

A. General

1. References: *Standard Methods* (1989, pp. 10-3 through 10-15; pp. 10-31 through 10-35; pp. 10-48 through 10-54), Operator's Manual, Turner Fluorometer Model 111-003, Operator's Manual, Turner Spectrofluorometer Model 430.
2. Outline of Method:
   The fluorometric method is more sensitive than the spectrophotometric technique for the detection of chlorophyll *a*. The optimum sensitivity is obtained at an excitation wavelength of 430 nm and an emission wavelength of 663 nm. The chlorophyll is extracted into acetone and the resulting fluorescence measured. The fluorometer is calibrated with chlorophyll solutions of known concentrations. Chlorophyll *a* may be overestimated if pheophytin *a*, which fluoresces near the same wavelength, is present. Pheophytin *a* is a common degradation product of chlorophyll *a*; measurements for both pigments must be made in order to correct for the pheophytin concentration. The ratio of chlorophyll *a* to pheophytin *a* is a good indicator of the physiological condition of the phytoplankton. In the sample, acidification with dilute acid releases the magnesium atom from the chlorophyll *a* molecule, converting it to pheophytin *a*. The fluorescence is read before and after acidification and concentrations for both chlorophyll *a* and pheophytin *a* are calculated.
3. Extract the samples immediately after they are taken. If this is not possible, the samples must be concentrated, frozen, and kept in the dark. Store in this manner for no longer than 30 days.

B. Special Reagents and Equipment

1. *Acetone, 100%*: Use analytical reagent grade or redistilled.
2. *Acetone, 90% v/v Aqueous Solution*: To 1800 mL of reagent grade or redistilled acetone in a glass bottle, add 200 mL DDW and mix well.
3. *Magnesium Carbonate Suspension*: Add 1.0 g finely powdered $MgCO_3$ to 100 mL DDW.

4. *Stock Chlorophyll Solution*: Pure chlorophyll *a* standard is available from Sigma Chemical Co., St. Louis, Missouri. Protect it from light at all times and store in a freezer. Very carefully weigh the contents of a sealed 1 mg vial on a micro-balance capable of weighing to the nearest ng (1×10^9 g). Rinse a glass funnel and 250 mL volumetric flask very well with 90% acetone. Cover the entire flask with aluminum foil. Using the funnel, transfer the chlorophyll to the flask quantitatively. Rinse the funnel well into the flask and dilute to volume with 90% acetone. Store in a glass container protected from light (covered with aluminum foil, black tape, etc.) (See 6.2 C below.) Store in a freezer.
4. *Hydrochloric Acid, 1N*: Add 5 mL of concentrated HCl to 55 mL of DDW. Store in a dropper bottle.
5. *Tissue Grinder*: Use a Teflon® or glass grinder, preferably having a grinding tube and pestle with rounded bottoms.
6. *Centrifuge and Centrifuge Tubes*: Use 15 mL graduated tubes with screw caps, chemically resistant to acetone.
7. *Glass Fiber Filters*: Use Whatman GF/C or the equivalent.
8. *Sonicator*: This should be equipped with a probe to dismember and homogenize the sample.
9. *Fluorometer or Spectrofluorometer*: *Turner Model 111-003* or the equivalent should be equipped with high intensity blue lamp and envelope (F4T.5), red sensitive photomultiplier tube (R-136), sliding windows (1X, 3X, 10X, 30X) and a high sensitivity door. *Turner Model 430* or equivalent should be equipped with an emission monochromator to set wave length range and bandwidth, a polarizer to reduce the effect of scattered light, a xenon lamp as a light source, and range settings of X1000, X300, X100, X30, X10, and X3.
10. *Cuvettes*: Use round-bottomed 12×75 mm borosilicate cuvettes.

    Note: Use only those clean and free of scratches.

11. *Filters*:

    Turner Model 111-003: Use excitation (primary filter), 5-60 (very dark purple); use emission (secondary filter), 2-64 (dark red).

    Note: Other filters may be substituted. Check operator's manuals.

    Turner Model 430: Use a polarizing filter mounted in cardboard; use an emission filter, 2A (pale yellow). Neutral density filters are optional: nominal transmittances, 10%, 1%, etc.

C. Standardization

1. Concentration of stock standard (approximately 4 mg/L)

Calculate the concentration of the stock standard using the spectrophotometric chlorophyll *a* procedure. Acidify the sample and check for the presence of pheophytin *a*. The before-to-after acidification ratio (OD 664_b/OD 665_a) should be close to 1.70 since the fresh stock standard contains very little pheophytin. Alternately (or in addition), calculate the concentration of the stock standard using the weight of the pure chlorophyll *a* weighed out (see calculations). The stock concentrations calculated by the two methods should vary by no more than 10% for accurate work.

Record the mass of chlorophyll weighed, the concentration(s) calculated, and the method(s) used to determine the concentration.

2. Daily Standardization
   a. Repeat the procedure outlined above. Discard the stock when it becomes degraded or contaminated with pheophytin. The best results are achieved with a healthy chlorophyll stock.
   b. Prepare a series of dilutions with 90% acetone from the stock standard and determine the fluorescence of each as outlined in the procedure. Protect the standards from light at all times. Calculate the concentration of each standard (see Calculations).

      Use standards that fluoresce in the same range as the samples. In order to produce a good linear standard curve, it will be necessary to read at least five standards on each sensitivity range (or window) of the fluorometer.
   c. Use the neutral density filters if it is necessary to read very concentrated samples (see Notes). In this case, better results are achieved if the fluorescence of all the standards and samples are determined using the same neutral density filter (10%, 1%, etc).
3. Quality Control
   Chlorophyll analytical procedures are subject to many problems not easily detected. Especially at the lower concentration levels, it is advisable to analyze a quality control sample also.

D. Procedure

Note: All glassware and cuvettes should be clean and acid-free.

Plankton Samples: Free-floating or suspended phytoplankton and/or zooplankton

1. Collect water samples using Van Dorn or Kemmerer water samplers. (See *Standard Methods* (1989, pp. 10-4 through 10-15) for further dis-

cussion.) If the water is oligotrophic or phytoplankton densities are expected to be low, collect 6 L of sample. Eutrophic waters will require a volume of 1 to 2 L. Record the depth at which the sample was taken.
2. Filter a known quantity of water through a glass fiber filter to concentrate the sample. Add 0.2 mL magnesium carbonate suspension in the final phase of filtering as a preservative. Store these concentrated samples frozen in a desiccator in the dark if acetone extraction is not possible immediately.
3. Place the filter in a tissue grinder tube and cover with 2 to 3 mL of 90% acetone. Macerate well without losing any sample.
4. Transfer the filter and solution to a centrifuge tube. Using 90% acetone, rinse the grinding tube and pestle into the centrifuge tube. Adjust the total volume to 5.0 mL (or some other known volume up to 10 mL) with 90% acetone. Allow the solution to stand overnight at 4°C in the dark.
5. Clarify the extract by centrifuging in closed tubes for 20 minutes at 500 g. Pour the supernatant into a calibrated centrifuge tube. Measure and record the total volume of the extract. Take care to keep the supernatant extract as clear as possible and free from suspended pieces of filter paper. Wrap the tubes with aluminum foil to protect them from light.
6. Refer to the appropriate fluorometric procedure below.

Periphyton Samples: Attached or somewhat motile microscopic organisms associated with the surfaces of submerged objects

1. Scrape samples from the natural or artificial substrates in the body of water. (See *Standard Methods* (1989, pp. 10-48 through 10-51) for a further discussion.) Add DDW to the sample bottle to bring it up to a known total volume (usually 100 mL). (It is helpful to calibrate the sample bottle at the 100 mL mark prior to sampling.) Record the surface area of the region scraped in in^2 or m^2 and the final total volume. Add 0.2 mL of magnesium carbonate suspension as a preservative. Store these samples frozen in the dark if acetone extraction is not possible immediately.
2. Homogenize the sample by sonication or other suitable means.
3. Withdraw a sample aliquot of known volume to another container. Mix well with an appropriate volume of 100% (not 90%) acetone to make the final solution 90% acetone (example: 10 mL of homogenized sample plus 90 mL of acetone).

   Note: If the sample is difficult to homogenize, it may be necessary to perform replicate extractions or increase the size of the sample aliquot.

If a large quantity of silt or nonchlorophyll material has been collected with the sample, the periphyton can be centrifuged and the supernatant extracted.

4. Allow the solution to stand overnight at 4°C in the dark.
5. Clarify the extract by centrifuging (as above). Alternately, withdraw the sample carefully from the supernatant to perform relative fluorescence measurements. Avoid agitation of settled material in the extract. Keep the sample bottle or centrifuge tube covered with aluminum foil to minimize light degradation.
6. Refer to the appropriate fluorometric procedure below.

In Vivo Samples: Example: Algal Bioassays

1. Swirl to homogenize the culture.
2. Withdraw enough sample to fill a cuvette. It is not necessary to use a sterile cuvette, but do not pour any nonsterile solution back into the culture.
3. Measure the fluorescence of the sample using the appropriate procedure below. Use DDW to zero the fluorometer instead of 90% acetone. Measurements can be converted to the appropriate amount of chlorophyll or biomass present.

Turner Model 111-003

1. Open the door and insert the primary filter (5-60) on the right side of the instrument. Slide in the secondary filter (2-64) on the left side.
2. Turn on the power to the instrument and then the light source. Open the door and make certain the light source is on by looking through the primary filter. Allow about five minutes for the instrument to warm up.

   Note: Emissions from the blue light source can be harmful to the eye. Do not look directly into the lamp for any length of time.

3. Fill a cuvette with 90% acetone for use as a blank. It is important to clean the outside of the cuvette thoroughly for blanks and samples.
4. Place the blank in the sample compartment and set the fluorometer dial to zero using the blank knob.
5. Measure the relative fluorescence of the standards and samples. Change the window (30X, 10X, 3X) by sliding the window orifice as necessary. Avoid any readings that are less than 20 or greater than 80 relative fluorescence units. Record relative fluorescence and the window (30X, 10X, 3X) used.

Note 1: Use the 1X window only for indication of dilutions needed. Results in this range will not be linear due to quenching effects.

Note 2: Make sure that each sample is read using a window where five standards have been read also.

Note 3: Rezero the fluorometer often and every time the window is changed.

6. Acidify the samples and standards in the cuvettes one at a time with 2 drops of 1N HCl. Gently agitate.
7. Record the fluorescence not sooner than 1 minute or later than 2 minutes after acidification. Treat all samples identically.

   Note: The fluorescence should always decrease after acidification.

8. Turn the power switch to off and clean the sample compartment.
9. Pour all extracted samples and waste acetone into a labeled container for repurification by distillation.

Turner Model 430

1. Set the wavelength counters on the right side of the instrument to 440 nm for excitation and 670 nm for emission.
2. Install emission filters on the left side of the sample compartment. The polarizing filter (cardboard) goes closest to the machine. Then slide in the 2A (pale yellow) filter.
3. If necessary, check the bandwidth setting under the operating console. (See Operator's Manual.) Both slide adjustments should be set on 60 nm.
4. If the instrument is used frequently, leave the power switch in the warm-up position.
5. Turn the power switch to the on position.
6. Turn on the xenon lamp by briefly depressing the xenon start button. There will be a crackling noise as the lamp lights. Make sure the lamp is on by checking for a glow in the lamp chimney. Do not look directly at the lamp without eye protection.
7. Allow 10 minutes to warm up if the power switch has been left in the warm-up position. If the instrument was in the off position, allow 1 hour warm-up before operating.
8. Fill a cuvette with 90% acetone as a blank. It is essential that the inside and outside of the cuvettes are scrupulously clean for samples and blanks. Wipe the outside dry with a Kimwipe.
9. Insert the blank in the sample holder and set the range switch to X1000.

10. Depress the shutter control to the read position (or to the record position for continuous operation).
11. Adjust the blank know until the meter reads zero.

    Note 1: Begin all readings with the meter damp switch set to 0.2 for a faster meter response. Higher range switch settings (X1000, X300) will require further damping (1 or 5) for a more stable final reading. This control can be changed without altering the actual readings.

    Note 2: The setting for the sensitivity control will vary with instrument conditions and the range of concentration studied. Consult the Operator's Manual or previous operators for suggested settings. Record the sensitivity setting for all measurements made. Only one sensitivity setting can be used within a given position of the range switch and a given set of samples and standards within that range.

12. Pour the standards into cuvettes and read their relative fluorescence, recording the range switch setting also.

    Note 1: Use the X1 setting only for indications of dilutions needed. Results in this range will not be linear due to quenching effects.

    Note 2: Rezero the instrument often and every time the range setting is changed.

    Note 3: Read relative fluorescence on the 0–100 meter scale when using range settings of X10, X100, or X1000. Read on the 0–30 meter scale when using range settings of X3, X30, or X300.

    Note 4: Read a minimum of five standards on each range switch setting used. Readings must be between 20 and 80 units on the 0–100 meter scale and between 6 and 26 units on the 0–33 meter scale.

    Note 5: Protect all standards and samples from light at all times. Degradation is rapid when the chlorophyll concentration is very low.

13. Read the relative fluorescence of the samples in the same manner as the standards.
14. Acidify the samples and standards in the cuvettes one at a time with 2 drops of 1N HC1. Gently agitate.
15. Record the fluorescence not sooner than 1 min or later than 2 min after acidification. Treat all samples identically.
16. Remove the last sample and clean the sample compartment.
17. Turn the power switch to warm-up, or to off if the instrument will not be used in the next few days.
18. Record the number of hours the xenon lamp was on. The xenon lamp must be replaced after 1500 hours or when lamp performance becomes erratic (consult operator's manual).

19. Pour all extracted samples and waste acetone into a labelled container for repurification by distillation.

E. Calculations

1. Chlorophyll *a* in stock standard:
   Calculate by one or both of the following methods:

   a. $$\text{mg Chl } a/\text{L} = \frac{M}{0.25} = \text{C (below)} \approx 4 \text{ mg/L or } \mu\text{g/mL}$$

   where
   M = mass of chlorophyll *a* weighed in mg recorded to the nearest ng.

   b. See 6.1.

2. Chlorophyll *a* in standards prepared daily:

$$\mu\text{g Chl } a/\text{L} = \frac{V_1 \times C}{V_2}$$

where
V_1 = volume of stock chlorophyll standard in mL.
C = concentration of stock standard in μg/mL (above).
V_2 = final volume of the diluted stock standard in L.

Note: Dilutions of the concentrated stock may become too large to produce accurate standards in the low μg/L range. Use one or more of the standards (concentrations calculated above) to prepare further dilutions. C in the above equation will then become the concentration of the freshly diluted standard from the stock.

3. Chlorophyll *a* in the extract without correction for pheophytin *a*:
   Using the relative fluorescence values for the standards and blank for each window or range setting, graph concentration (μg/L) versus relative fluorescence to obtain a straight line. Read the concentrations of the samples from the graph or use a calculator with a linear regression program.
4. Chlorophyll *a* in the extract with the correction for pheophytin *a*.
   a. Within each range (sensitivity, s), calculate a calibration factor (F_s) for each standard:

$$F_s = \frac{C_a}{R_s}$$

where

C_a = μg Chl *a*/L for each standard
R_s = relative fluorescence units for each standard.

Then compile all data for the standards within each range and calculate an average calibration factor, $\bar{F}_s$.

b. Within each range (sensitivity, s) calculate a before-to-after acidification ratio (r) for each standard:

$$r = \frac{r_b}{r_a} \approx 1.7 \text{ if chlorophyll stock is healthy}$$

where

r_b = relative fluorescence of each standard before acidification
r_a = relative fluorescence of each standard after acidification

Then compile all data for the standards within each range and calculate a mean value before-to-after acidification ratio, $\bar{r}$.

c. Calculate the "corrected" chlorophyll *a* and pheophytin *a* as follows:

$$\mu g \text{ Chl } a/L = \bar{F}_s \left(\frac{\bar{r}}{\bar{r} - 1}\right)(R_b - R_a)$$

$$\mu g \text{ Pheo } a/L = \bar{F}_s \left(\frac{\bar{r}}{\bar{r} - 1}\right)(\bar{r}R_a - R_b)$$

where

$\bar{F}_s$, $\bar{r}$ are as calculated above.

R_b = relative fluorescence of the sample extract before acidification.
R_a = relative fluorescence of the sample extract after acidification.

5. Chlorophyll concentrations in plankton samples:

$$\mu g \text{ Chl } A/L = \frac{\mu g \text{ Chl } a/L \times A}{B}$$

where

mg Chl A/L = concentration of chlorophyll a in the extract as determined above.

A = total volume in liters of the extract supernatant after centrifuging.

B = volume of the sample filtered in liters.

(Similar calculations can be completed for pheophytin also.)

6. Chlorophyll concentrations in periphyton samples:

$$\mu g \text{ Chl } A/m^2 = \mu g \text{ Chl } a/L \times 10 \times C/D$$

where

mg Chl A = concentration of chlorophyll a in the extract as determined above.

C = total sample volume in the field situation in liters; usually 0.1 L.

D = surface area in m^2

Note: $m^2 = in.^2 \div 1550$

(Similar calculations can be completed for pheophytin also.)

F. Notes

1. Neutral density filters can be used to reduce the total amount of light transmitted and detected by the photomultiplier. This can allow the analyst to work in a higher concentration range without doing dilutions. These filters do not change color characteristics but do decrease sensitivity. They are marked with their nominal transmittance (10%, 1%, etc.) and should be inserted on the secondary side of the instrument with the emission filter for best results. As an example, the 10% filter will allow only 10% of the light through; therefore, final results must be multiplied by 10. However, more accurate results can be obtained by preparing standards that are higher in concentration and then reading standards and samples using the same filter.

2. If the fluorometer is used frequently, data from standards can be compiled and for each range a factor can be calculated for direct conversion of relative fluorescence to chlorophyll *a*. However, these factors should be reevaluated frequently by preparation of fresh chlorophyll stock and the analysis of quality controls.

6.3 Total Coliform Membrane Filter Method

A. General

1. References: *Standard Methods* (1989, pp. 9–82 through 9–91) and EPA (1978, pp. 32–58).
2. Outline of Method: The sample is obtained in a sterile container and is filtered through a sterile 0.45 μm membrane filter. The filter is then placed on a sterile pad saturated with liquid media and is incubated at 35 ± 0.5°C for 22–24 hours. Sheen colonies are counted under low magnification. The enrichment procedure is suggested for assessing drinking water.

   The total coliform group includes all of the aerobic and facultative anaerobic, gram-negative, nonspore-forming, rod-shaped bacteria that ferment lactose in 24–48 hr at 35°C. This includes the following genera: *Escherichia, Citrobacter, Enterobacter,* and *Klebsiella.*
3. Bacteriological samples cannot be preserved. Analyze samples preferably within six hours of collection. Otherwise, analyze within 24 hours of collection.

B. Materials and Culture Media

1. *Sample Bottles*: Pyrex glass wide mouth bottle with rubber lined cap, approximately 125 mL capacity. Whirl-Pak bags are acceptable. For larger filtration volumes, any screw cap bottle capable of being sterilized is acceptable.
2. *Petri Dishes*: 50 mm, sterile, disposal plastic, tight sealing. Relative humidity of 90% must be maintained inside the dish. Humidity may be maintained by incubation inside any container in which the atmosphere is saturated with water.
3. *Pipets and Pipet Containers*: Pre-sterilized bacteriological serological 1 mL and 10 mL pipets are recommended. Otherwise, pipets can be wrapped in paper or aluminum or stainless steel containers for sterilization.
4. *Filtration Assembly*: Standard Millipore-type filter holder assembly (glass, porcelain, or any noncorrosive bacteriologically inert metal) with side arm filtration flask.
5. *Filter Membranes*: 0.45 μm Millipore filters, pre-sterilized.

6. *Absorbent Pads*: Sterile Millipore pads.
7. *Forceps*: Nonserated with round or blunt tips.
8. *Microscope*: A binocular wide field dissecting scope; a small fluorescent lamp with magnifier is acceptable. Optical system with an incandescent light source is unsatisfactory.
9. *Incubator*: Capable of temperature 35 ± 0.5°C.
10. *Culture Media*: Dissolve 4.8 g dehydrated M-Endo Broth MF in 100 mL DDW containing 2 mL of ethanol. Heat to boiling. Don't prolong boiling or heat in autoclave. Cool to room temperature, covered with foil. Store in the dark at 2-10°C and discard any unused media after 96 hours.
11. *Enrichment Culture Media*: Lauryl typtose broth. Make up according to the labeled instructions provided by the manufacturer.
12. *Phosphate Buffer Stock*: Dissolve 34.0 g KH_2PO_4 in 500 mL DDW, adjust pH to 7.2 with 1N NaOH, and dilute to 1 L with DDW. Store in refrigerator.
13. *Magnesium Chloride Solution*: Dissolve 38 g $MgCl_2$ in DDW and dilute to 1 L. Store in refrigerator.
14. *Dilution Water*: Prepare one of the following:
    a. *Buffered Water*: Add 1.25 mL phosphate buffer stock solution and 5.0 mL magnesium chloride solution to 1 L DDW. Dispense in amounts that will provide 99 ± 2.0 mL after autoclaving for 15 min.
    b. *Peptone Water*: Add 1 g peptone to 1 L DDW. Final pH should be 6.8. Dispense in amounts that will provide 99 ± 2.0 mL after autoclaving for 15 min.
15. *Sterilization*: Sample bottles, petri dishes (other than plastic), filtration units, and graduated cylinders are sterilized in an autoclave at 121°C (15–17 psi) for 15 min. Pipets and containers are sterilized in an oven at 170°C for 2 hours. Also, filtration units and graduated cylinders may be sterilized by placing them in an UV cabinet for 2 min.

    CAUTION: Always use slow exhaust when liquids are included in the material to be sterilized in the autoclave.

C. Standardization

No standardization, as such, is run. However, positive and negative controls should be set up. A positive control is a check on the quality of the media; a negative control is a check on the sterile technique used.

Positive Control: Use a pure culture that will give a positive reaction with the media. Set up a positive control for each batch of media that is prepared. Follow the technique outlined below for filtering and incubation; set up appropriate dilutions so that the plates are within the countable range of 20–80 colonies per plate.

Negative Control: Use sterile dilution water as a negative (sterile) control. Set up a negative control for each batch of media and for each series of samples. Follow the technique below for filtering and incubation. If the negative control plates show colonies (contamination), disregard data from that series of samples. The waters involved must be resampled and analyzed again.

D. Procedure

1. Sampling: Collect sample in a sterile bottle such that water **doesn't** flow across your hand and into bottle. In a moving stream, point opening of bottle upstream and sweep bottle through water against the current, again, so there is no hand contamination. Leave an air space in the container to facilitate mixing the sample just prior to filtration.
2. Alcohol-flame forceps prior to each handling of filters or pads. Touch only the edge of the filters with the forceps. Use one sterile filtration apparatus for each sample.
3. Select sample aliquots or dilutions based upon expected bacterial density, about 50 coliform colonies and not more than 200 colonies of all types. It is standard procedure to use at least three different aliquot sizes or dilutions per sample, in order to cover a wide range of bacteriological concentrations. If an aliquot of less than 20 mL is filtered, a small amount of sterile dilution water should be added to the funnel **before** filtration.
4. Place pad in petri dish bottom (smaller diameter half) and pipet 1.8-2.0 mL of M-Endo Broth MF on absorbent pad. (Use enough broth to saturate the pad. Excess media may be poured out of the dish.)
5. Vigorously shake sample for 30 seconds. Place membrane filter in filtration apparatus using sterile technique.
6. Measure the sample with a sterile pipet or a sterile graduated cylinder; filter the sample under partial vacuum.
7. Rinse filter using three 20-30 mL portions of sterile dilution water.

   Note: Within a dilution series, filter in order from the highest dilution (smallest amount of sample) to lowest dilution (greatest amount of sample).

8. Remove membrane filter from filtration unit and roll it on the pad, avoiding any entrapment of air. Replace dish half.
9. Invert petri dishes and incubate for 22–24 hours at 35 ± 0.5°C.
10. Count typical coliform colonies, i.e., those having a pink to dark red color with a golden-green metallic surface sheen when viewed with the fluorescent light source.
11. Compute coliform densities from the membrane filter count within the 20–80 coliform colony range.
12. If the number of colonies is outside the range of 20–80 colonies and still countable, calculate the total coliforms/100 mL and note that the count is estimated.

E. Calculation

$$\text{Total Coliform/100 mL} = \frac{\text{coliform colonies counted} \times 100}{\text{mL sample filtered}}$$

When reporting the results, always give the method used, i.e., membrane filter method.

F. Enrichment Procedure

1. Filter sample as above.
2. Place pad on petri dish top (larger diameter half) and pipet 1.8–2.0 mL of enrichment media on the pad.
3. Remove membrane filter from filtration unit and roll it on the pad, avoiding entrapment of air. Replace dish half.
4. Incubate the filter, without inverting the dish, for 1½–2 hours at 35 ± 5°C in an atmosphere of at least 90% relative humidity.
5. Remove from incubator and separate dish halves. Place a fresh sterile pad on the bottom half of the petri dish and saturate it with 1.8–2.0 mL of M-Endo Broth MF.
6. Roll membrane filter off enrichment pad (discard old pad) and roll it on the new M-Endo pad. Replace dish half.
7. Incubate the dish, inverted, for 20–22 hours at 35 ± 0.5°C.
8. Count and compute as above.

   Note: After tests have been completed, disposable petri dishes are autoclaved at 121°C (15–17 psi) for 15 min. This will destroy the dish (melt it) and any bacteria present. Remains may then be thrown away.

6.4 Fecal Coliform
Membrane Filter Method

A. General

1. References: *Standard Methods* (1989, pp. 9–82 through 9–84; 9–94 through 9–96) and EPA (1978, pp. 32–58).
2. Outline of Method:
   The sample is obtained in a sterile container and is filtered through a sterile 0.45 μm membrane filter. The filter is then placed on a sterile pad saturated with liquid media and is incubated at 44.5 ± 0.2° for 22–24 hours. Blue Colonies are counted under low magnification.

   Fecal coliforms are part of the total coliform group. Fecal coliforms are defined as gram-negative, nonspore-forming, rod-shaped bacteria that ferment lactose in 24 ± 2 hours at 44.5 ± 0.2°c. The major species is *Escherichia coli*.
3. Bacteriological samples cannot be preserved. Analyze samples preferably within six hours of collection. Otherwise analyze within 24 hours of collection.

B. Materials and Culture Media

1. *Samples Bottles*: Pyrex glass wide mouth bottles with rubber lined cap, approximately 125 mL capacity. Whirl-Pak bags are acceptable. For larger filtration volumes, any screw cap bottle capable of being sterilized is acceptable.
2. *Petri Dishes*: 50 mm, sterile, disposal plastic, tight sealing. Relative humidity of 90% must be maintained inside the dish. Humidity may be maintained by incubation inside any container in which the atmosphere is saturated with water.
3. *Pipets and Pipet Containers*: Pre-sterilized bacteriological serological 1 mL and 10 mL pipets are recommended. Otherwise, pipets can be wrapped in paper or aluminum or stainless steel containers for sterilization.
4. *Filtration Assembly*: Standard Millipore filter holder assembly (glass, porcelain or any noncorrosive bacteriologically inert metal) with side arm filtration flask.
5. *Filter Membranes*: 0.45 μ Millipore filters, pre-sterilized.
6. *Absorbent Pads*: Use sterile Millipore pads.

7. *Forceps*: Nonserrated with round or blunt tips.
8. *Microscope*: A binocular wide field dissecting scope; a magnifier with a light source is acceptable.
9. *Incubator*: Stirring water bath capable of 44.5 ± 0.2°C temperature tolerance.
10. *Plastic Bags*: Use Whirl-Pak bags, for incubation.
11. *M-FC Broth*: Dissolve 3.7 g of broth in 100 mL DDW. Add 1.0 mL of 1% rosolic acid in 0.2N NaOH. Heat to boiling; promptly remove from heat and cool to below 45°C. DO NOT AUTOCLAVE. Media can be stored at 2–10°C. The unused portion must be discarded 96 hours after preparation. Rosolic acid is stable for one week. Discard if it appears brownish in color.
12. *Phosphate Buffer Stock*: Dissolve 34.0 g KH_2PO_4 in 500 mL DDW, adjust pH to 7.2 with 1N NaOH, and dilute to 1 L with DDW. Store in a refrigerator.
13. *Magnesium Chloride Solution*: Dissolve 38 g $MgCl_2$ in DDW and dilute to 1 L. Store in refrigerator.
14. *Dilution Water*: Prepare one of the following:
    a. *Buffered Water*: Add 1.25 mL phosphate buffer stock solution and 5.0 mL magnesium chloride solution to 1 L DDW. Dispense in amounts that will provide 99 ± 2.0 mL after autoclaving for 15 min.
    b. *Peptone Water*: Add 1 g peptone to 1 L DDW. Final pH should be 6.8. Dispense in amounts that will provide 99 ± 2.0 mL after autoclaving for 15 min.
15. *Sterilization*: Sample bottles, petri dishes (other than plastic), filtration units, and graduated cylinders are sterilized in an autoclave at 121°C (15–17 psi) for 15 min. Pipets and containers are sterilized in an oven at 170°C for two hours. Also, filtration units and graduated cylinders may be sterilized by placing them in a UV cabinet for 2 min.

    CAUTION: Always use slow exhaust when liquids are included in the material to be sterilized in the autoclave.

C. Standardization

See Total Coliform.

D. Procedure

1. Sampling: Collect sample in a sterile bottle such that water **doesn't** flow across your hand and into the bottle. In a moving stream, point opening of bottle upstream and sweep bottle through water against the current, again, so there is no hand contamination. Leave an air space in the container to facilitate mixing the sample just prior to filtration.
2. Alcohol-flame forceps prior to each handling of filters or pads. Touch only the edge of the filters with the forceps. Use one sterile filtration apparatus for each sample.
3. Select sample aliquots or dilutions based upon expected bacterial density, about 20 to 60 colonies. It is standard procedure to use at least three different aliquot sizes or dilutions per sample, in order to cover a wide range of bacteriological concentrations. If an aliquot of less than 20 mL is filtered, a small amount of sterile dilution water should be added to the funnel **before** filtration.
4. Place absorbent pad in petri dish and pipet approximately 2 mL of M-FC medium to saturate the pad. Excess media may be poured out of the dish.
5. Filter as in Total Coliform.
6. Roll membrane filter on pad being careful to exclude air bubbles. Replace dish half.
7. Wrap 4 to 6 dishes in a Whirl-Pak bag and anchor below the water surface with dishes inverted and horizontal. (Anchor well below surface to maintain critical temperature requirements).
8. Incubate in water bath for 24 hours at $44.5 \pm 0.2°C$.
9. Count colonies produced by fecal coliform bacteria; they are blue in color. Nonfecal coliform colonies are gray to cream color. Desired fecal coliform range: 20–60 colonies per filter.
10. Colony counts outside the 20–60 range may be calculated for fecal coliform/100 mL; mark these calculations estimated.

E. Calculation

$$\text{Fecal Coliform/100 mL} = \frac{\text{fecal coliform colonies counted} \times 100}{\text{mL sample filtered}}$$

When reporting the results, always give the method used, i.e., membrane filter method.

Note: After tests have been completed, disposable petri dishes are autoclaved at 121°C (15–17 psi) for 15 min. This will destroy the dish (melt it) and any bacteria present. Remains may then be thrown away.

6.5 Heterotrophic (Standard) Plate Count Method

A. General

1. References: *Standard Methods* (1989, pp. 9–54 through 9–64) and EPA (1978, pp. 101–107).
2. Outline of Method: The sample is obtained in a sterile container and is pipetted into a sterile plastic petri dish. Liquified plate count agar medium is then added. Other agars such as R2A and NWRI can also be used. The plate is rotated to evenly distribute the bacteria and is then incubated at 35 ± 0.5°C for 48 hours. The cultures are counted and reported as colony-forming units per mL or CFU/mL.
3. Bacteriological samples cannot be preserved. Analyze samples preferably within 6 hours of collection. Otherwise, analyze within 24 hours of collection.

B. Materials and Culture Media

1. *Sample Bottles*: Pyrex glass wide mouth bottle with rubber lined cap, approximately 125 mL capacity.
2. *Petri Dishes*: Sterile 100 mm × 15 mm petri dishes, glass, or plastic.
3. *Sterile Bacteriological Pipets*: Glass or plastic of appropriate volumes, usually 1.0 mL in 0.1 graduations.
4. *Incubator*: Capable of maintaining a stable 35 ± 0.5°C. Temperature is checked against an NBS certified thermometer or one of equivalent accuracy.
5. *Water Bath*: For tempering agar set at 44–46°C.
6. *Colony Counter*: Quebec Dark-Field model or equivalent.
7. *Hand Tally*: Or electronic counting device (optional).
8. *Bunsen/Fisher Gas Burner*: Or electric incinerator.
9. *Sterile Plate Count Agar*: Suspend 23.5 g tryptone glucose yeast agar in 1 L of DDW; heat to boiling to dissolve the agar; autoclave to sterilize. The pH of the medium at 25°C should be 7.0. If preparing pour plates, allow the medium to cool to 44–46°C before pouring. To facilitate cooling, immerse flask into a water bath incubator until plates are ready to be poured.
10. *Sterile Peptone Dilution Water*: Add 1 g of Bacto-peptone to 1 L of DDW. Dispense into dilution bottles (approximately 105 mL to obtain 99 mL ± 2 after autoclave exposure) and sterilize by autoclaving. Final pH should be 6.8–7.0.

11. *Dilution Bottles*: (Milk dilution), pyrex glass, marked at 99 mL volume, screw cap with neoprene rubber liner.
12. *Sterilization*: Sample bottles and petri dishes (other than plastic) are sterilized in an autoclave at 121°C (15–17 psi) for 15 min.

C. Standardization

No standardization, as such, is run. However, positive and negative controls should be set up. A positive control is a check on the quality of the media; a negative control is a check on the sterile technique used.

Positive Control: Use a pure culture that will give a positive reaction with the medium. Set up appropriate dilutions so the plates are within the countable range of 30–300 colonies per plate.

Negative Control: Use sterile dilution water as a negative control. If the negative control plates show colonies (contamination), disregard data from that series of samples. The waters involved must be resampled and analyzed again.

D. Procedure

1. Sampling: See Total Coliforms.
2. Dilutions: The desired final plate count is 30–300 colonies. If the microbial population in the sample is not known, a series of dilutions must be prepared to obtain a count within this range. See Figure 6.1 for making up dilutions.
3. Prepare duplicate plates for each sample or dilution tested. Vigorously shake the sample or dilution before transferring to the petri dishes. Using aseptic technique, pipet an aliquot of each dilution into the bottom of the respective petri dish. Use a different sterile pipet for each dilution and sample. After delivery, touch the tip of the pipet once to a dry spot.
4. Add approximately 15 mL (not less than 12 mL) of cooled (44–45°C) plate count agar medium to each petri dish. Mix the inoculated medium by slowly tracing the plate through a figure eight pattern.
5. After the agar has hardened (usually within 10 min) invert and incubate at 35 ± 0.5°C for 48 ± 3 hours.
6. After the incubation period, count the colonies using a Quebec-type colony counter or use a hand tally to aid in counting the colonies.
7. Plates with 30 to 300 colonies are calculated and reported as Heterotrophic (standard) plate count per mL or colony-forming units per mL

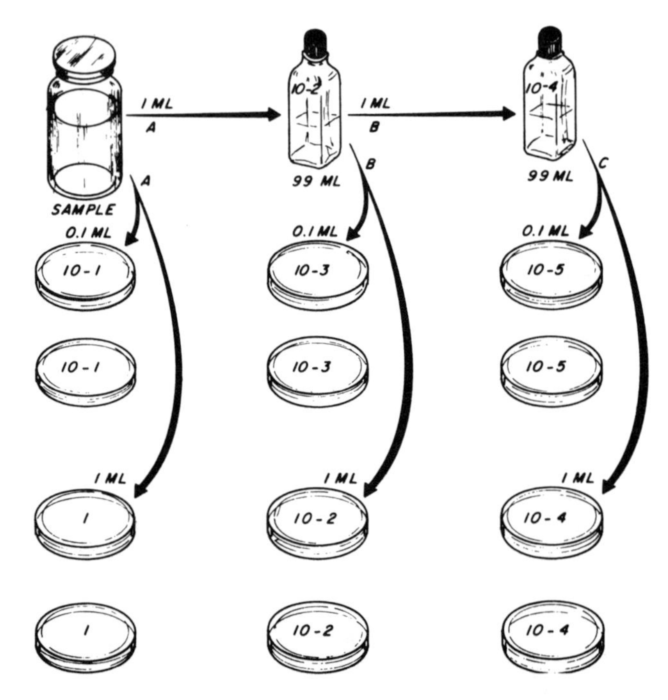

Figure 6.1. Typical dilution series for standard plate count (EPA, 1978).

(CFU/mL). Plates with counts less than 30 are reported as estimated CFU/mL. Plates with greater than 300 colonies are reported as a greater than (>) CFU/mL.

8. See EPA (1978, pp. 101-107) for an extended discussion of reporting results and for any difficulty encountered with reading the plates.

E. Calculation

Heterotrophic (Standard) Plate Count, CFU/mL = average number of colonies per plate × the recriprocal of the dilution used

When reporting the results in CFU/mL, always include the method used, incubation temperature, time, and medium used.

Note: After tests have been completed, disposable petri dishes are autoclaved at 121°C (15–17 psi) for 15 min. This will destroy the dish (it melts) and any bacteria present. Remains may then be thrown away.

Appendix A: Precision and Accuracy Forms

Precision Data and Statement.

Quality Control Data; Precision

Parameter: ____________________ Date: ________

Method: ____________________ Analyst: ________

Reference: ____________________

concentration (μg, mg/L)

n	Sample #	Sample #	Sample #	Sample #
1				
2				
3				
4				
5				
6				
7				
8				
9				
10				
$\bar{x}$				
s				
RSD				

Notes: Table 1.10 outlines precision terms and calculations needed above. *Precision Statement:* Using water samples at concentrations of ________ μg, mg/L and ________ μg,mg/L NH_3-N, the standard deviations were ________ and ________ units, respectively. The relative standard deviations were ________ % and ________ %, respectively.

Accuracy Data and Statement.

Quality Control Data; Accuracy

Parameter: ______________________ Method: ______________________

Reference: ______________________ Analyst: ______________________ Date: __________

Sample #: ______ UNSP = ______ (μg, mg)/L CONSP = ______ (μg, mg)/L TH-SP = ______ (μg, mg)/L				Sample #: ______ UNSP = ______ (μg, mg)/L CONSP = ______ (μg, mg)/L TH-SP = ______ (μg, mg)/L			Sample #: ______ UNSP = ______ (μg, mg)/L CONSP = ______ (μg, mg)/L TH-SP = ______ (μg, mg)/L			Sample #: ______ UNSP = ______ (μg, mg)/L CONSP = ______ (μg, mg)/L TH-SP = ______ (μg, mg)/L		
n	EX-SP	%D	%R	EX-SP	%D	%R	EX-SP	%D	%R	EX-SP	%D	%R
1												
2												
3												
4												
5												
6												
7												
8												
9												
10												
$\overline{\%R}$												
s												
RSD												

Notes: This form is applicable to spiking procedures for analyses listed in Tables 1.1, 1.2, 1.3, 1.6, and 1.7. Calculate TH-SP by consulting these tables. Table 1.10 outlines other accuracy terms and calculations needed above. S_1 = ________; S_2 = ________; S_3 = ________.

Accuracy Statement: Using water samples at concentrations of ________ μg, mg/L and ________ μg, mg/L, ________, the recoveries were ________% and ________%, respectively. The standard deviations for percent recoveries were ± ________ and ± ________ units, respectively. The relative standard deviations were ________% and ________%, respectively.

Accuracy Data and Statement.

Quality Control Data; Accuracy

Parameter: ____________ Method: ____________

Reference: ____________ Analyst: ____________ Date: ____________

Spiking Compound: ____________ S_1: ____________

	Sample #: ____ UNSP = ____ (μg, mg)/L Approx. M ____ mg QC = ____ (μg, mg)/L				Sample #: ____ UNSP = ____ (μg, mg)/L Approx. M ____ mg QC = ____ (μg, mg)/L				Sample #: ____ UNSP = ____ (μg, mg)/L Approx. M ____ mg QC = ____ (μg, mg)/L				Sample #: ____ UNSP = ____ (μg, mg)/L Approx. M ____ mg QC = ____ (μg, mg)/L			
n	M	TH-SP	EX-SP	%R	M	TH-SP	EX-SP	%R	M	TH-SP	EX-SP	%R	M	TH-SP	EX-SP	%R
1																
2																
3																
4																
5																
6																
7																
8																
9																
10																
$\overline{\%R}$																
s																
RSD																

Notes: This form is applicable to spiking procedures for analyses listed in Tables 1.4, 1.5, and 1.8 Calculate TH-SP by consulting these tables. When using a spiking compound, TH-SP values will be different for each replicate spiked sample. Table 1.10 outlines other accuracy terms and calculations needed above. *Accuracy Statement:* Using water samples at concentrations of ________ μg, mg/L and ________ μg, mg/L ________, the recoveries were ________% and ________%, respectively. The standard deviations for percent recoveries were ± ________ and ± ________ units, respectively. The relative standard deviations were ________% and ________%, respectively.

Appendix B: Atomic Weights

International Atomic Weights (^{12}C = 12.000 amu)

Element	Symbol	Atomic Number	Atomic Weight	Element	Symbol	Atomic Number	Atomic Weight
Actinium	Ac	89	(227)	Mercury	Hg	80	200.59
Aluminum	Al	13	26.9815	Molybdenum	Mo	42	95.94
Americium	Am	95	(243)	Neodymium	Nd	60	144.24
Antimony	Sb	51	121.75	Neon	Ne	10	20.183
Argon	Ar	18	39.948	Neptunium	Np	93	(237)
Arsenic	As	33	74.9216	Nickel	Ni	28	58.70
Astatine	At	85	(210)	Niobium	Nb	41	92.906
Barium	Ba	56	137.34	Nitrogen	N	7	14.0067
Berkelium	Bk	97	(247)	Nobelium	No	102	(254)
Beryllium	Be	4	9.0122	Osmium	Os	76	190.2
Bismuth	Bi	83	208.980	Oxygen	O	8	15.9994
Boron	B	5	10.811	Palladium	Pd	46	106.4
Bromine	Br	35	79.909	Phosphorus	P	15	30.9738
Cadmium	Cd	48	112.40	Platinum	Pt	78	195.09
Calcium	Ca	20	40.08	Plutonium	Pu	94	(244)
Californium	Cf	98	(249)	Polonium	Po	84	(210)
Carbon	C	6	12.01115	Potassium	K	19	39.102
Cerium	Ce	58	140.12	Praseodymium	Pr	59	140.907
Cesium	Cs	55	132.905	Promethium	Pm	61	(145)
Chlorine	Cl	17	35.453	Protactinium	Pa	91	(231)
Chromium	Cr	24	51.996	Radium	Ra	88	(226)
Cobalt	Co	27	58.9332	Radon	Rn	86	(222)
Copper	Cu	29	63.54	Rhenium	Re	75	186.2
Curium	Cm	96	(245)	Rhodium	Rh	45	102.905
Dysprosium	Dy	66	162.50	Rubidium	Rb	37	85.47
Einsteinium	Es	99	(254)	Ruthenium	Ru	44	101.07
Erbium	Er	68	167.26	Samarium	Sm	62	150.35
Europium	Eu	63	151.96	Scandium	Sc	21	44.956
Fermium	Fm	100	(252)	Selenium	Se	34	78.96
Fluorine	F	9	18.9984	Silicon	Si	14	28.086
Francium	Fr	87	(223)	Silver	Ag	47	107.870
Gadolinium	Gd	64	157.25	Sodium	Na	11	22.9898
Galllium	Ga	31	69.72	Strontium	Sr	38	87.62
Germanium	Ge	32	72.59	Sulfur	S	16	32.064
Gold	Au	79	196.967	Tantalum	Ta	73	180.948
Hafnium	Hf	72	178.49	Technetium	Tc	43	(99)
Helium	He	2	4.0026	Tellurium	Te	52	127.60
Holmium	Ho	67	164.930	Terbium	Tb	65	158.924
Hydrogen	H	1	1.00797	Thallium	Tl	81	204.37
Indium	In	49	114.82	Thorium	Th	90	232.038
Iodine	I	53	126.9044	Thulium	Tm	69	168.934
Iridium	Ir	77	192.2	Tin	Sn	50	118.69
Iron	Fe	26	55.847	Titanium	Ti	22	47.90
Krypton	Kr	36	83.80	Tungsten	W	74	183.85
Lanthanum	La	57	138.91	Uranium	U	92	238.03
Lawrencium	Lw	103	(257)	Vanadium	V	23	50.942
Lead	Pb	82	207.19	Xenon	Xe	54	131.30
Lithium	Li	3	6.942	Ytterbium	Yb	70	173.04
Lutetium	Lu	71	174.97	Yttrium	Y	39	88.905

Element	Symbol	Atomic Number	Atomic Weight	Element	Symbol	Atomic Number	Atomic Weight
Magnesium	Mg	12	24.312	Zinc	Zn	30	65.37
Manganese	Mn	25	54.9380	Zirconium	Zr	40	91.22
Mendelevium	Mv	101	(256)				

The value given in parentheses denotes the mass number of the longest-lived or best-known isotope.

Definitions of Abbreviations, Terms, and Units

amu atomic mass units

BOD biochemical oxygen demand

COD chemical oxygen demand

DDW Doubly deionized water. This is water used for all blanks, reagents, standards, dilutions, and rinsing. The water passes through standard cation-anion exchange columns and then through another set of laboratory cation-anion exchange columns. This second purification system also includes an activated carbon column to remove organic impurities. DDW may also be referred to as "laboratory pure water" or "reagent grade water." The resistance of DDW should be 10 to 18 megohms/cm. When the resistance decreases to ≤ 7 megohms/cm, the ion exchange columns need to be replaced.

df dilution factor

dilutions

1. 1 → 10. This means one part diluted up to 10 parts, usually with DDW.

   Example: 1.0 mL standard or sample plus 9.0 mL DDW for a final volume of 10.0 mL.

   Example: 10.0 mL standard or sample in a 100 mL volumetric flask diluted up to the 100 mL mark.

2. 1 + 10. This means 1 part plus 10 parts. The final volume here would be 11 parts.

   Example: 1 mL H_2SO_4 plus 10 mL DDW.

   Note: Any time a standard or sample is diluted, it should be done accurately and volumetrically.

DO dissolved oxygen

Dry to Constant Weight Applies to gravimetric analyses when a tared, preweighed container or filter is required. Dry the weighing container or the container plus sample at 103°C for at least 1 hr. Cool in a desiccator and weigh on an analytical balance to the nearest

	0.1 mg. Repeat this process until the weights differ by no more than ±°0.5 mg or ±4% of the previous weight, whichever is less.
EDTA	ethylenediaminetetraacetic acid or its salts
eq	equivalent(s)
FAS	ferrous ammonium sulfate
g	gram(s)
K	specific conductance or conductivity
kg	kilogram(s)
KHP	potassium hydrogen phthalate
L	liter(s)
LAS solution	linear alkylbenzene sulfonate solution
M	mole or molar
μg	microgram(s), 1×10^{-6} g
mg	milligram(s), 1×10^{-3} g
μg/L	Micrograms per liter; 1×10^{-6} g per liter. This is approximately equivalent to "parts per billion" (ppb).
mg/L	Milligrams per liter; 1×10^{-3} g per liter. This is approximately equivalent to "parts per million" (ppm).
μL	microliter(s)
mL	milliliter(s)
μmhos/cm	General unit for reporting specific conductance or conductivity—micromhos per centimeter.
N	normal
ng	nanogram(s), 1×10^{-9} g
NH_3-N, NO_3^--N, PO_4^{-3}-P, etc	Describes a concentration or a mass of a compound expressed as the element that follows the compound name. For example, 50 μg NH_3-N/L means 50 μg/L ammonia expressed

as nitrogen mass rather than the mass of the whole compound. Similarly, 4 mg NO_3^--N/L means 4 mg/L nitrate expressed as nitrogen.

nm nanometers(s)

NTU Nephelometric turbidity unit(s)

pH $-\text{Log}[H^+]$

psi pounds per square inch

sp. gr. specific gravity

TC To contain. This is usually found in small print on volumetric flasks. The volume specified on the flask is accurate to ±1% when the liquid is at 20°C and the meniscus is at the line. When the liquid is poured out of the flask, it will not be exactly the volume specified, since some of the liquid will adhere to the sides of the glass.

TD To deliver. This is commonly found in small print on pipets, particularly volumetric pipets. This means that the volume specified on the pipet will be accurate to usually ±1%, when the liquid is at 20°C and the meniscus is at the line marked on the pipet. Do *not* blow out the small amount of liquid left in the tip of the pipet; the volume is calibrated with this taken into account. Touch the pipet tip to the inside of the glassware into which it is draining. This will ensure deliverance of an accurate volume. Allow the pipet to drain by gravity; do not blow into the pipet to hurry the process.

V/V Volume-to-volume. This describes the composition of a solution. For example, a 25% v/v solution of sulfuric acid is made by adding 250 mL conc H_2SO_4 to 750 mL DDW. For sulfuric acid, the resulting 25% v/v solution is nearly 25% H_2SO_4, since concentrated sulfuric acid contains 96–98% of the active ingredient. However, for other solutions made from a concentrated stock, this will not always be the case. For example, a 25% v/v solution of hydrochloric acid is made by adding 250 mL conc HCl to 750 mL DDW; conc HCl is 36–37% of the active ingredient. Therefore, the resulting 25% v/v solution of hydrochloric acid will be less than 10% HCl.

Bibliography

Included here are references cited in the procedures and other references pertaining to water and environmental chemistry, analytical procedures for water quality, and related topics.

American Water Works Association. 1958. *Safety Practices for Water Utilities.* Manual M6. American Water Works Association, New York, NY.

Barnes, H. 1959. *Apparatus and Methods of Oceanography. Part 1: Chemical.* Interscience, NY.

Bennett, G. F., F. S. Feates, and I. Wilder. 1982. *Hazardous Materials Spills Handbook.* McGraw-Hill Book Co., New York, NY.

Creitz, G. L., and F. A. Richards. 1955. The estimation and characterization of plankton populations by pigment analyses. III. A note on the use of Millipore membrane filters in the estimation of plankton pigments. *J. Mar. Res.* 14:211–216.

Deberry, D. W., J. R. Kidwell, and D. A. Malish. 1982. *Corrosion in Potable Water Systems.* U.S. Environmental Protection Agency, Office of Drinking Water, (WH-550), Washington, DC.

Dux, J. P., and R. F. Stalzer. 1988. *Managing Safety in the Chemical Laboratory.* Van Nostrand Reinhold Co., New York, NY.

EPA. 1976. *Quality Criteria for Water.* U.S. Environmental Protection Agency, Washington, DC.

EPA. 1978. *Microbial Methods for Monitoring the Environment; Water and Wastes.* U.S. Environmental Protection Agency, EPA 690/8–7S-017, Cincinnati, OH.

EPA. 1979. *Handbook for Analytical Quality Control in Water and Wastewater Laboratories.* U.S. Environmental Protection Agency, EPA-600/4–79–019, Cincinnati, OH.

EPA. 1982. *Handbook for Sampling and Sample Preservation of Water and Wastewater.* U.S. Environmental Protection Agency, EPA-600/4–82–029, Cincinnati, OH.

EPA. 1983. *Methods for Chemical Analysis of Water and Wastes.* U.S. Environmental Protection Agency, EPA-690/4–79–020, Cincinnati, OH.

EPA. 1986. *Quality Criteria for Water 1986.* U.S. Environmental Protection Agency, Office of Water Regulations and Standards, EPA 440/5–86–001, Washington, DC.

EPA. 1987. *Handbook of Methods for Acid Deposition Studies, Laboratory Analysis of Surface Water Chemistry.* U.S. Environmental Protection Agency, Acid Deposition and Atmospheric Research Division, Office of Acid Deposition, Environmental Monitoring, and Quality Assurance,

Office of Research and Development, EPA 600/4-87/026, Washington, DC.

Federal Register. 1978. Vol. 43(45): March 7. As cited in Oceanography International Corporation. December 15, 1979. *Standard Ampule Method and Low COD Value Ampule Method*. College Station, TX.

Federal Register. 1979. Interim primary drinking water regulations: control of trihalomethanes in drinking water. Vol. 44(231): November 29.

Hazards in the Chemical Laboratory. 1983. L. Bretherick, Ed. The Royal Society of Chemistry, London.

Hunt, D. T. E., and A. L. Wilson. 1986. *The Chemical Analysis of Water*. The Royal Society of Chemistry, Burlington House, London.

Inhorn, S. L. 1978. *Quality Assurance Practices for Health Laboratories*. American Public Health Association, Washington, DC.

Jardim, W. F. and J. J. R. Rohwedder. 1989. Chemical oxygen demand (COD) using microwave digestion. *Water Res*. 23:1069–1071.

Kenkel, J. 1988. *Analytical Chemistry for Technicians*. Lewis Publishers, Inc., Chelsea, MI.

Kopp, J. F., and R. C. Kramer. 1967. *Trace Metals in Waters of the United States*. U.S. Department of the Interior, FWPCA, Division of Pollution Surveillance, Cincinnati, OH.

Leischman, A. A., J. C. Greene, and W. E. Miller. 1979. *Bibliography of Literature Pertaining to the Genus* Selenastrum. EPA-600/9-79-021, Corvallis, OR.

Manahan, S. E. 1990. *Environmental Chemistry*. 4th edition. Lewis Publishers, Inc., Chelsea, MI.

Manufacturing Chemists' Association, General Safety Committee. 1972. *Guide for Safety in the Chemical Laboratory*. 2nd edition. C. Van Nostrand Co., New York, NY.

McKee, J. E., and H. W. Wolf. 1963. *Water Quality Criteria*. 2nd edition. Publication 3-A. California State Water Quality Control Board, Sacramento, CA.

Menzel, D. W., and R. F. Vaccaro. 1964. The measurement of dissolved organic and particulate carbon in seawater. *Limnol. Oceanogr*. 9:138–142.

Miller, W. E., J. C. Greene, and T. Shiroyama. 1978. *The* Selanastrum capricornutum *Printz Algal Assay Bottle Test: Experimental Design, Application, and Data Interpretation Protocol*. EPA-600/9-78-013, Corvallis, OR.

National Research Council. 1981. *Prudent Practices for Handling Hazardous Chemicals in the Laboratory*. National Academy Press, Washington, DC.

National Research Council. 1983. *Prudent Practices for Disposal of Chemicals from Laboratories*. National Academy Press, Washington, DC.

Parker, C. R. 1972. *Water Analysis by Atomic Absorption Spectroscopy*. Varian Instruments Division, Sunnyvale, CA.

Parsons, T. R., and J. D. H. Strickland. 1963. Discussion of spectrophotometric determination of marine plant pigments, with revised equations for ascertaining chlorophylls and carotenoids. *J. Mar. Res.* 21:155–163.

Parsons, T. R., Y. Maita, and C. M. Lalli. 1984. *A Manual of Chemical and Biological Methods for Seawater Analysis.* Pergamon Press, Elmsford, NY.

Phifer, R. W., and W. R. McTigue, Jr. 1988. *Handbook of Hazardous Waste Management for Small Quantity Generators.* Lewis Publishers, Chelsea, MI.

Pitts, M. E., and V. D. Adams. 1987. Method for total nitrogen in freshwater and wastewater samples. In *Proceedings of the AWWA 1986 Water Technology Conference, Advances in Water Analysis and Treatment*, pp. 849–858. American Water Works Association, Denver, CO.

Principles of Environmental Sampling. 1988. L. H. Keith, Ed. American Chemical Society, Washington, DC.

Quality Assurance Manual. 1979. Environmental Chemistry, Bureau of Laboratories, Utah State Division of Health, Salt Lake City, UT.

Richards, F. A., and T. G. Thompson. 1952. The estimation and characterization of plankton populations by pigment analyses, II. A spectrophotometric method for the estimation of plankton pigments. *J. Mar Res.* 11:156–172.

Safe Drinking Water Committee. 1977. *Drinking Water and Health.* National Academy of Sciences, Washington, DC.

Safe Handling of Cryogenic Liquids. 1987. Compressed Gas Association, Inc., Arlington, VA.

Safe Storage of Laboratory Chemicals. 1984. D. A. Pipitone, Ed. John Wiley & Sons, New York, NY.

Saunders, G. N., F. B. Trama, and R. W. Bachman. 1962. *Evaluation of a Modified* ^{14}C *Technique for Shipboard Estimation of Photosynthesis in Large Lakes.* Great Lakes Research Division., University of Michigan Publication No. 8:1–61. Ann Arbor, MI.

Sawyer, C. N., and P. L. McCarty. 1978. *Chemistry for Environmental Engineering.* McGraw-Hill, New York, NY.

Schmalz, K. L. 1971. *Phosphorus Distribution in the Bottom Sediments of Hyrum Reservoir.* Master's Thesis, Utah State University, Logan, UT.

Singley, J. E., B. A. Beaudet, and P. H. Markey. 1984. *Corrosion Manual for Internal Corrosion of Water Distribution Systems.* U.S. Environmental Protection Agency, Office of Drinking Water, Washington, DC.

Slovacek, R. E., and P. J. Hannan. 1977. In vivo fluorescence determinations of phytoplankton chlorophyll *a*. *Limnol. Oceanogr.* 22(5):919–924.

Snoeyink, V. L., and D. Jenkins. 1980. *Water Chemistry.* John Wiley & Sons. New York, NY.

Solorzano, L. 1969. Determination of ammonia in natural waters by the phenolhypochlorite method. *Limnol. Oceanogr.* 14(5):799–801.

Solorzano, L., and J. H. Sharp. 1980. Determination of total dissolved nitrogen in natural waters. *Limnol. Oceanogr.* 25(4):751–754.

Standard Methods for the Examination of Water and Wastewater. 1989. 17th ed. American Public Health Association, Washington, DC.

Strickland, T. R., and J. D. H. Parsons. 1972. *A Practical Handbook of Seawater Analysis*. Fisheries Research Board of Canada, Ottawa.

Taylor, J. K. 1987. *Quality Assurance of Chemical Measurements*. Lewis Publishers, Chelsea, MI.

Technicon Industrial Systems. 1978. *Industrial Methods, Nitrate and Nitrite in Water and Wastewater (Range: 0.04–2.0 mg/N/L)*. No. 102-7OW/B, Tarrytown, NY.

Turner Associates. 1977. *Operator's Manual for Turner Model 111-003 Fluorometer*. Sequoia Turner Corporation, Mountain View, CA.

Turner Associates. 1978. *Operator's Manual for Turner Model 430 Spectrofluorometer*. Sequoia Turner Corporation, Mountain View, CA.

Walkley, A. 1935. An examination of the methods for determining organic carbon and nitrogen in soils. *J. Agr. Sci.* 25:598–609.

Walters, D. B. 1980. *Safe Handling of Chemical Carcinogens, Mutagens, Tratogens, and Highly Toxic Substances*. Vols. 1 and 2. Ann Arbor Science, Ann Arbor, MI.

Water Analysis. 1988. W. Fresenius, K. E. Quentin, and W. Schneider, Eds. Springer-Verlag Berun Heidelberg, New York, NY.

Young, J. C., G. N. McDermott, and D. Jenkins. 1981. Alterations in the BOD procedure for the 15th edition of *Standard Methods for the Examination of Water and Wastewater*. *J. Water Pollut. Control Fed.* 53:1253–1259.

Zadorojny, C., S. Saxton, and P. Finger. 1973. Spectrophotometric determination of ammonia. *J. Water Pollut. Control Fed.* 45:905–912.

Index